Food Processing and Production: Principles and Applications

Food Processing and Production: Principles and Applications

Kaden Hunt

Larsen & Keller
www.larsen-keller.com

Food Processing and Production: Principles and Applications
Kaden Hunt
ISBN: 978-1-64172-096-0 (Hardback)

Larsen & Keller

Published by Larsen and Keller Education,
5 Penn Plaza,
19th Floor,
New York, NY 10001, USA

Cataloging-in-Publication Data

Food processing and production : principles and applications / Kaden Hunt.
    p. cm.
Includes bibliographical references and index.
ISBN 978-1-64172-096-0
1. Food industry and trade. 2. Produce trade. 3. Food supply. 4. Food industry and trade--Appropriate technology.
I. Hunt, Kaden.
TP370 .F66 2019
664--dc23

For more information regarding Larsen and Keller Education and its products, please visit the publisher's website www.larsen-keller.com

Table of Contents

Preface

The process of transformation of agricultural products into food for consumption is termed as food processing. It is an important aspect of food production. It is necessary for making food edible, enabling preservation, toxin removal, and distribution and marketing. This is achieved through a number of processes, such as food preservation, use of food additives, and food packaging. With processing, delicate perishable food can be transported across longer distances and used for longer periods of time as it kills pathogenic microbes and deactivates spoilage. A variety of traditional and industrial methods are used for food preservation, such as freezing, pickling, heating, pasteurization, vacuum packing, irradiation, etc. Food additives can be added to the food during the manufacturing process, processing and packaging, or during storage and transport to preserve its taste and flavor. These may include antioxidants, emulsifiers, acidity regulators, colors, etc. Packaging ensures that food becomes tamper-proof and stays protected from various biological and chemical agents. This book is a compilation of chapters that discuss the most vital concepts in the field of food processing and production. It unfolds innovative principles and applications, which will be crucial for the holistic understanding of the subject matter. For all those who are interested in this field, this textbook will prove to be an essential guide.

A foreword of all Chapters of the book is provided below:

Chapter 1, Modern food production is characterized by innovative and sophisticated technologies and methods in the areas of preservation, processing and storage. This is an introductory chapter, which will introduce briefly all the significant topics related to the production of food, such as aging, convenience food, shelf-stable food, processing aid, frozen food, etc.; **Chapter 2**, The transformation of food by various chemical and physical methods into edible food products is called food processing. It involves a number of activities such as liquefaction, mincing, macerating, emulsification and preservation. This chapter has been carefully written to provide an easy understanding of the varied processes of food processing; **Chapter 3**, Food preservation is an important process in food production. Its objective is to prevent the growth of microorganism for slowing the oxidation of fats that are responsible for rancidity of food. The topics elaborated in this chapter will help in providing a better perspective about the different food preservation techniques, such as curing, freezing, boiling, pickling, canning, pasteurization, etc. It also elucidates the modern practices such as the use of nonthermal plasma, high-pressure food preservation, biopreservation, pulsed electric field electroporation, etc.; **Chapter 4**, A food additive is a substance that is used in food to preserve and enhance its appearance, flavor and taste. All the varied food additives and nutraceuticals as well as the processes of food fortification, food coloring and food coating have been covered in extensive detail in this chapter; **Chapter 5**, Science and technology have undergone rapid developments in the past decade, which has resulted in innovation in industrial methods of food production such as snap freezing, ultra-high temperature processing, dry milling and fractionation of grain, Ohmic heating, etc. These have been extensively discussed in this chapter. It also explores the principles of basic industrial processes like food drying, brining and extrusion for a comprehensive understanding of the field.

I would like to thank the entire editorial team who made sincere efforts for this book and my family who supported me in my efforts of working on this book. I take this opportunity to thank all those who have been a guiding force throughout my life.

Kaden Hunt

Introduction to Food Production

Modern food production is characterized by innovative and sophisticated technologies and methods in the areas of preservation, processing and storage. This is an introductory chapter, which will introduce briefly all the significant topics related to the production of food, such as aging, convenience food, shelf-stable food, processing aid, frozen food, etc.

Food production, as the name suggests, is all about preparing food in which raw materials are converted into ready-made food products for human use either in the home or in the food processing industries. It process comprises of art and scientific approach. Food production has many sections and it starts with basic things like cleaning, packing, segregating, sorting, preparing, adding ingredients in correct proportions, presenting, etc.

Process of Food Production

There are large numbers of plant and animal products, which are used for our wellbeing. They provide us food and the daily food, which we eat, comes from both plants and animals. These include grains, pulses, spices, honey, nuts, cereals, milk, vegetables, fruits, egg, meat, chicken, etc. The existence of our life is completely depended on plants and animals. Altogether both plant and animal species provide 90% of the global energy.

Types of Food Production

Food production is further classified into different types including, cultivation, selection, crop management, harvesting, crop production, preserving, baking, pasteurizing, pudding, carving, butchers, fermenting, pickling, drink and candy makers, restaurants, etc.

Methods of Food Production

- Chopping or slicing of vegetables.

- Curing food.

- Grinding and marinating.

- Emulsification.

- Food fermentation.

- Fermenting beer at brewing industries.

- Boiling, broiling, frying, grilling, steaming and mixing.

- Pasteurization.

- Fruit juice processing.

- Removing the outer layers either by peeling and skinning.

- Gasification of soft drinks.

- Preserving and packaging of food products by vacuum packs.

Aging

The process of maturing a food or beverage to improve the flavor of the item is known as ageing. The purpose of storing a product so it can age is to improve the overall taste and to impart the distinct characteristics provided by each storing technique. For wine and many liquors, ageing refers to the amount of time the product is stored in oak barrels, stainless steel barrels, or in bottles. Beverages can be aged for a few weeks or for lengths of time that can be for many decades.

When applied to cheeses or meats, ageing decreases the amount of moisture contained in the food, which results in a shaper and more concentrated flavor. Typically, when cheese or meat is aged, the food also becomes harder in texture. Cheeses become less buttery or smooth textured while the color darkens. Meats can become grainer in texture and noticeably less moist.

Drying

Drying of foods by leaving them in a low-humidity environment has been used as a food preservation technique for millennia. Air-dried meat such as jerky may have been some of the first preserved foods ever eaten by man. Drying also concentrates flavors in foods by removing water from them.

Fermentation

Foods may be aged to allow fermentation to occur, such as in the making of alcoholic beverages, in cheesemaking, in pickling, such as kimchi, and in meat or fish products such as fermented sausage or surstromming.

Culturing

Besides fermentation, microbial food cultures can act on food products to alter their chemical make-up and provide additional flavors. This is especially true in processes such as the making of blue cheese or aged beef.

Extraction

In the case of beverages, such as the aging of wine, beer, or whiskey, storing the beverage for extended periods of time in wooden casks allows the liquid to extract flavor compounds from the wood itself, adding to the complexity and depth of flavor. Traditional Balsamic Vinegar is aged for years in a succession of oak barrels to extract and concentrate flavors.

Ready Meals

Pre-prepared food is big business in a modern society where demanding work and social activities have put the squeeze on meal times. For ready meal producers, product integrity and an appetising look is key to success in this market. Consumers demand first class appearance, even spread of ingredients and product integrity and, according to many producers, a 'smashed up' ready meal is the biggest driver of lost custom.

A ready meal line can be made up of many processing and packaging options as the diverse range of ingredients, either fresh or frozen, are handled and then presented to the consumer. In terms of processing, pumps can be deployed for operations such as transfer and mixing to produce powerful suction with low shear, low pulsation and gentle handling for such products as mashed potato. Also, depositors have to cope with filling multi-component meals and deposit precise portions of product into trays, cartons or pouches. Browning grills can then be used to give a baked-like finish to ready meals.

In terms of packaging, the tray has been one of the most popular formats for ready meals which use either a modified atmosphere or gas flushing on fresh ingredients. Once the trays have been filled, they are sealed and then tested for any leaks to ensure product integrity at all times. Many meals are then placed in cartons to provide optimum shelf impact.

In recent times, there has been a plethora of other packaging formats used for 'fresh meal kits' including pouches and rigid plastics pots but the traditional, more high volume ready meal continues to be packed in trays.

A line can compose of tray denesters that feed trays to single or twin lane conveyors and arrive at a volumetric filler which can be used for bulky ingredients such as rice and pasta. Filled trays can then move to another filling station which applies a pre-mixed sauce.

Trays are then vibrated, to help settle the contents, before moving on to the distribution system below a multi head weigher, through which the most costly ingredients are generally added. These are often the main protein content of the meal such as meat, poultry or seafood.

Trays can then pass on to other ingredient adding stations or straight to the tray sealer, where they are sealed under vacuum before then being check weighed. Often, trays then proceed to an auto-clave for retorting to provide a shelf life of three months.

Convenience Food

Convenience food is any item of food that is prepared before you purchase it. It also involves some kind of packaging and made available in food stores. There are certain preservatives, chemicals and ingredients added to them to enable them to last longer because it might take a long time before anyone buys them. In other words, they are prepared for prolonged shelf life but this might mean they are not as nutritious as you would have hoped.

Convenient foods have become widely popular especially among working class people, teenage children, people living in hostels, bachelors, sharing rooms etc. Convenience foods are used to shorten the time of meal preparation at home. Some convenient foods can be eaten immediately or after adding some water, heating or thawing.

Some popular, easy to prepare convenient foods are: Masala Oats, Corn flakes, canned soup, frozen foods (sausages, ham, bacon etc), bread etc. Other convenience foods are cake mixes, spice mix powder, sauces etc that are pre-cooked and sold. Most of the convenience foods takes hardly less than 5 minutes to cook the food. They are often prepared or packaged before being consumed and can be used at any time, quickly and easily by thawing or heating the food.

It is a fact that convenient foods are designed to be cheap, tasty and non-perishable but the ingredients added into these packaged products contain added sugars and fats. Trans-fat is typically used as they don't spoil the food and corn syrup is a cheap way to make the product sweet. Salt can also add awesome flavor very inexpensively to the food. Most of the convenience foods have become very popular because they can be served as a quickie snack or meal.

Convenience foods may offer some fantastic such as less time spent in the kitchen or planning meals, less preparation time, fewer leftovers and easy cleaning up.

Advantages of Convenience Foods

- Preparation time is reduced to a great extent.

- No storing, buying or planning of ingredients.

- Can hardly get any leftovers.

- Could have a variety of items especially for inexperienced cooks.

- Faster presentation and easy cleaning up.

- Less spoilage and waste occur with packaged convenience foods.

- Transportation of packaged foods is cheaper especially in concentrated form.

- Cost efficient for mass production and distribution.

- Ready to eat cereal and instant breakfast difficult to prepare at home because of its expensive product technology used in preparation.

Disadvantages of Convenience Foods

- May be less meat, fish, or cheese than you would include in homemade versions.

- Cooking time is sometimes increased for thawing or longer baking time.

- Harder to control fat, salt and sugar levels.

- Cost per serving may be higher than homemade.

- Convenience foods are typically high in calories, fat, saturated fat, sugar, salt, and trans-fats.

- They tend to lack freshness in fruits and vegetables.

Shelf-stable Foods

Shelf stable food is food of a type that would normally be stored refrigerated but which has been processed so that it can be safely stored in a sealed container at room or ambient temperature for a usefully long shelf life. For instance, the first shelf stable formulation of ranch dressing, created in 1983, had a shelf life of 150 days. Various food preservation and packaging techniques are used to extend a food's shelf life. Decreasing the amount of available water in a product, increasing its acidity, or irradiating or otherwise sterilizing the food and then sealing it in an air-tight container, can all extend a food's shelf life without unacceptably changing its taste or texture. For some foods alternative ingredients can be used. Common oils and fats become rancid relatively quickly if not refrigerated; replacing them with trans fats delays the onset of rancidity, increasing shelf life. This is a common approach in industrial food production, but recent concerns about health hazards associated with trans fats have led to their strict control in several jurisdictions. Even where trans fats are not prohibited, in many places there are new labeling laws, which require information to be printed on packages, or to be published elsewhere, about the amount of trans fat contained in certain products.

Shelf-Stable Foods

Here are some examples of what kinds of foods are considered shelf-stable:

Meat, Fish, and Protein- Canned Fish (Tuna, Salmon), Canned Chicken, Canned Ham, Beef or Turkey Jerky, Beef Sticks, Non-Refrigerated Pepperoni, Canned Beans (Lentils, Chickpeas, Kidney Beans, Black Beans), Canned refried beans, Nuts (Almonds, Walnuts, Peanuts, Cashews, Pecans etc.), Peanut Butter, Almond Butter.

Vegetables- Canned Green Beans, Canned Sweet Corn, Canned Creamed Corn, Canned Carrots, Canned Peas, Canned Tomatoes.

Liquids- Juices (canned, bottled, Individual Boxes), Stock (chicken, beef, vegetable).

Condiments (only considered shelf-stable until opened, after which they need to be refrigerated)-Ketchup, Mustard, Mayo, Salad Dressing, Oils, Vinegars, and Sauces, Salsa, Canned/Jarred Pickles/Relish/Olives.

Grains- Cereal, Cereal Bars, Crackers, Tortilla, Chips, Pretzels, Granola, Granola Bars, Protein Bars, Couscous, Grits and Instant, Oatmeal (Add room temp water, allow extra time for reconstituting).

Fruit- Canned Peaches, Canned Pears, Canned Pineapple, Canned Fruit Cocktail, Applesauce, Dried Plums, Dried Apples, Dried Apricots, Dried Raisins, Dried Cranberries, Dried Blueberries, Dried Banana Chips.

Dairy- Shelf stable (ultra-processed) Milk/Soymilk/Almond milk, Cheese in a Can (cheese wiz), Velveeta, Shelf-Stable Cheese-Dip.

Dry Herbs/Seasonings- Salt & Pepper, Cayenne Pepper, Cumin, Coriander, Curry, Garlic Powder, Onion Powder, Thyme, Rosemary, Oregano.

Frozen Food

The terms fresh frozen or frozen fresh can both be used to mean that the food was quickly frozen while it was still fresh. It seems that the term fresh frozen is more often used, and this is probably because it is most open to a wrong interpretation. That is, consumers are more likely to think that it implies freshness, instead of how quickly it was brought to the freezer. "Frozen fresh," on the other hand, will probably be interpreted as frozen while fresh. They both mean the same thing. The term freshly frozen can also be used.

Keep in mind that a fresh food can be blanched before freezing and still be called fresh frozen. This is because the blanching of some foods before freezing is not only common but required for certain foods, as a quick scald before freezing helps preserve nutrients.

Quickly frozen does not mean the same thing as fresh frozen or frozen fresh. This term, instead of describing how fresh the food was when it was frozen, describes the freezing method itself, and how fast it is. Particularly, it refers to freezing methods like blast-freezing, which uses sub-zero (in terms of Fahrenheit) temperatures and super-chilled air "blasted" at the food. The quicker a food is frozen, in many cases, the better, in terms of quality and deterioration, so quickly frozen is usually a plus.

Facts Related to Frozen Foods

1. Frozen foods do not require any added preservatives to keep them safe and consumable, because microbes—the kind that make you sick—cannot grow on any food that is at a temperature less than 0°F. The microbes don't die at that temperature, but they stop multiplying. Be careful when you unfreeze food; microbes will instantly start growing as they do on unfrozen food , so it's best to handle thawing food as you would fresh food.

2. Despite some old wives' tales, freezing food does not remove any nutrients. Freeze away.

3. You don't need to be afraid of freezer burn or color changes in your properly frozen food. Freezer burn is just the result of air hitting frozen food and allowing the ice to sublimate; other color changes can be blamed on long freezing times or poor packaging. It might look gross, but if your frozen food has maintained a proper temperature, it's fine to eat.

4. Freezing food typically keeps items edible indefinitely, although taste and quality may

diminish over time. Some items that stay tasty even after long freezes include uncooked game, poultry, and meat, which are still good even after up to a year in the freezer.

5. Even though freezing food was used as a storage technique in cold weather climates for many years, it's believed it was first applied to industrial food sales sometime in the 1800s, when a wily Russian company froze a small quantity of duck and geese and shipped them to London. By 1899, the Baerselman Bros. company adapted frozen storage for their own Russia-to-England food shipping business, though they initially only operated during cold weather months.

6. Carl Paul Gottfried Linde, an engineer, scientist, and professor at the Technical University of Munich, is basically the father of frozen food. Sort of. He helped pioneer industrial cooling, through what's commonly known as the Hampson-Linde cycle, and used his findings to plan an ice and refrigeration machine back in the nineteenth century.

7. Linde's desire to build such machines was only furthered in 1892, when the Guinness Brewery requested that Linde create a carbon dioxide liquefaction plant for them, pushing him still further into the arena of low temperature refrigeration and the liquefaction of air. Thanks, beer.

8. Ever wonder about the namesake of Birds Eye Frozen Foods? It came straight from the company's founder, Clarence Birdseye, who introduced the concept of flash freezing to the world.

9. Birdseye developed his technique after seeing food freezing in action in the Arctic, and noting how much better frozen fish tasted if it had been frozen immediately after been caught—like a flash!—versus food that was frozen on a delay.

10. Not only did Birdseye help pioneer flash freezing as a frozen food standard, he also helped develop in store freezer cases and refrigerated boxcars that allowed his frozen foods (and, yes, others) to travel near and far.

11. Birdseye's food was so prevalent that it was actually the first frozen food sold commercially in the United States. On March 6, 1930, Birds Eye frozen foods were put on sale at Davidson's Market in Springfield, Massachusetts, the first product of its kind.

12. The first "complete" frozen meal was not actually the beloved "TV dinner"—it was airplane food, In 1945, Maxson Food Systems, Inc. starting making their so-called "Strato-Plates," meals that were created specifically for consumption on airplanes (both by military and civilian passengers). Each frozen meal included a meat, a vegetable, and a potato, and was meant to be reheated for in-air chowing.

13. Maxson closed up shop before their Strato-Plates could be sold on the ground, but other companies picked up the slack, including One-Eyed Eskimo, Quaker State Food, and Swanson's, which is widely hailed as the true creator of TV dinners—they coined the name and were the most well-known maker of tasty compartmentalized meals in the 1950s.

14. Conagra Foods introduced its Healthy Choice line of frozen food back in 1989, after the

corporation was inspired to pursue healthy frozen picks after its chairman, Charles Harper, suffered a heart attack due to his bad eating habits.

15. There's long been a debate over which company first introduced the frozen pizza to the grocery store market, with both Totino's and Tombstone vying for the title. A more likely candidate? The Celentano brothers, who owned their own Italian specialty store in New Jersey in the 1950's, are believed to have marketed the first frozen pizza in 1957.

Processing Aid

Processing aids are substances or additives of natural or synthetic origin used in the production of foods. They are commonly used in a wide variety of products including bakery, confectionery, jams, jellies, meat and produce. The Food and Drug Administration or United States Department of Agriculture must approve processing aids prior to commercial use. They are considered extremely safe and are used in small volumes and do not alter the taste or appearance of the finished product.

The objective of this fact sheet is to describe the purpose and use of processing aids in the food industry.

Purpose of Processing Aids

The primary purpose of a processing aid is to facilitate the manufacturing of a food product. Processing aids are used for variety of reasons:

1. Improve product quality and consistency.

2. Enhance nutrition.

3. Help maintain product wholesomeness.

4. Enhance shelf life.

5. Help packing and transportation.

An incomplete list of process aids and their purposes is given in table.

Examples of Processing Aids

Processing aids can include everything from food contact lubricants used on equipment and pans to antimicrobials used in the final wash of produce to enhance shelf life and promote food safety. Other examples include foaming agent, pH regulator and anti-caking agent. Table gives examples of processing aids, food products they are used in and their purpose.

No.	Food Products	Processing Aid	Purpose
1	Apple juice	Gelatin with gums	Helps to eliminate suspended particles
2	Baked goods and baking mixes	Agar-agar	Vegan substitute for gelatin that helps the gelling of mixes
3	Beverages	Silicone	Antifoam
4	Bread	Phospholipase	Increase volume and prolongs softness
5	Cheese	Rennet	Separates curd and whey
6	Chill water	Ozone	Antimicrobial
7	Dough	Xylanase	Increases flexibility
8	Fish and meat (seafood)	Salt	Decrease water activity to improve shelf life
9	Frozen dough (e.g. waffles and pancakes)	Sodium sterol lactylate	Strengthens dough
10	Fruit and vegetable washes	Chlorine organic acid washes	Antimicrobial
11	Liquid nitrogen	BBQ sauce	Improves stability of plastic container
12	Meat	Ammonium hydroxide	Antimicrobial
13	Products transported on conveyors	Oil or synthetic	Lubricant
14	Sugar	Dimethylamine epichlorohydrin copolymer	Decoloring agent helps in clarification of sugar

Table: Examples and functions of processing aids incomplete list in alphabetical order according to product.

Criteria for Processing Aids

The criteria stipulated to qualify for a processing aid by the FDA are listed in. In the United States, the Food Safety Inspection Service determines if a substance meets the criteria for a processing aid. Canada does not have a regulatory definition of a food-processing aid; however, food additives require pre-clearance by the Canadian minister of health.

Criteria	No technical effect on final food product
Ingredients	No pre-approval process by FDA "Food Quality" under 21CFR Independent Evaluations + GRAS, EAFUS, FCC
Level	Used at level to obtain needed effect. Some chemicals may have specified max allowable levels. Trace levels may be present in final food.
Process Compliance	Good Manufacturing Practice

Table: FDA / United States Guidelines for Processing Aids

Ethics of using Processing Aids

Assuming processing aids are safe, ethical use is a top concern. Mehpam (2011) suggested three crucial ethical questions regarding food additives:

(1) Consumer sovereignty to act on informed judgments,

(2) Risk of harm to the consumer and

(3) The effects on laboratory animals during testing.

The answers to these questions may guide the ethical use of processing aids. Each concerned manufacturer and consumer should identify and explore the ethical issues associated with processing aids.

Ethical use of processing aids also may center around personal beliefs. Consumers on strict diets such as kosher, halal or vegetarian could have particular concerns about processing aids. For example, a vegetarian might want to avoid foods that have contacted processing aids made from animal fat.

Future of Processing Aids

Owing to the advantages of using processing aids, they are not likely to be eliminated; however, continuous improvements in processing methods and equipment may make them obsolete. Improvements in the formulation and application of processing aids also might make them more effective and more ethical for their intended use. Finally, companies may opt to select more ethical processing aids or include them in their ingredient list.

Nitrogen gas used as a processing aid in a barbecue sauce plant.

Processing aids are not required to be listed on the label, but some trace amounts of the material may remain in the product. Also, some processing aids are converted to normal constituents of the food but must not significantly increase the original amount. In any case, a processing aid is required to be "Generally Recognized as Safe" (GRAS). This means the overwhelming evidence considered by industry, academia and independent experts agrees the processing aid is safe for consumers.

Food-processing aids are important to the production of safe, quality foods. They perform valuable functions making them indispensable in many applications. Use of processing aids should be evaluated from the standpoint of food safety, ethics and efficiency (in that order) before use.

References

- Food-production, biology: byjus.com, Retrieved 21 April 2018

- What-are-convenience-foods: panlasangpinoy.com, Retrieved 11 May 2018

- Advantages-and-disadvantages-of-convenience-foods: vahrehvah.com, Retrieved 18 March 2018

- Michigan prepares, shelf-Stable-Foods-436127-7: michigan.gov, Retrieved 28 March 2018

- Food-law: what-is-fresh-frozen-food: culinarylore.com, Retrieved 08 May 2018

- 15-surprising-facts-about-frozen-food-62179: mentalfloss.com, Retrieved 18 July 2018

Food Processing

The transformation of food by various chemical and physical methods into edible food products is called food processing. It involves a number of activities such as liquefaction, mincing, macerating, emulsification and preservation. This chapter has been carefully written to provide an easy understanding of the varied processes of food processing.

Food processing is the set of methods and techniques used to transform raw ingredients into food or food into other forms for consumption by humans or animals either in the home or by the food processing industry. Food processing typically takes clean, harvested crops or slaughtered and butchered animal products and uses these to produce attractive, marketable, and often long-life food products. Similar processes are used to produce animal feed. Extreme examples of food processing include the expert removal of toxic portions of the fugu fish or preparing space food for consumption under zero gravity.

Examples of some processed foods.

The benefits of food processing include the preservation, distribution, and marketing of food, protection from pathogenic microbes and toxic substances, year-round availability of many food items, and ease of preparation by the consumer. On the other hand, food processing can lower the nutritional value of foods, and processed foods may include additives (such as colorings, flavorings, and preservatives) that may have adverse health effects.

Food processing dates back to prehistoric ages, with crude processing methods that included slaughtering, fermenting, sun drying, preserving with salt, and various means of cooking (such as roasting, smoking, steaming, and oven baking). Salt-preservation was especially common for foods that constituted the diets of warriors and sailors, up until the introduction of canning methods. Evidence for the existence of these methods exists in the writings of the ancient Greek, Chaldean, Egyptian, and Roman civilizations, as well as archaeological evidence from

Europe, North and South America, and Asia. These tried and tested processing techniques remained essentially the same until the advent of the Industrial Revolution. Examples of ready-meals also exist from the period before the Industrial Revolution, such as the Cornish pasty and Haggis.

Modern food processing technology was largely developed in the nineteenth and twentieth centuries, to serve military needs. In 1809, Nicolas Appert invented a vacuum bottling technique that would supply food for French troops, and this contributed to the development of tinning and then canning by Peter Durand in 1810. Although initially expensive and somewhat hazardous due to the lead used in cans, canned goods later became a staple around the world. Pasteurization, discovered by Louis Pasteur in 1862, was a significant advance in ensuring the microbiological safety of food.

In the twentieth century, World War II, the space race, and the rising consumer society in developed countries (including the United States) contributed to the growth of food processing with such advances as spray drying, juice concentrates, freeze drying, and the introduction of artificial sweeteners, coloring agents, and preservatives such as sodium benzoate. In the late twentieth century, products such as dried instant soups, reconstituted fruits and juices, and self-cooking meals (such as "Meal, Ready-to-Eat," or MRE, field rations) were developed.

In Western Europe and North America, the second half of the twentieth century witnessed a rise in the pursuit of convenience, as food processors marketed their products especially to middle-class working wives and mothers. Frozen foods found their success in the sales of juice concentrates and "TV dinners." Processors utilized the perceived value of time to appeal to the postwar population, and this same appeal contributes to the success of convenience foods today.

Food Processing Methods

Beer fermenting at a brewery.

Common food processing techniques include:

- Removal of unwanted outer layers, such as potato peeling or the skinning of peaches;
- Chopping or slicing, such as to produce diced carrots;
- Mincing and macerating;

- Liquefaction, such as to produce fruit juice;

- Fermentation, as in beer breweries;

- Emulsification;

- Cooking, by methods such as baking, boiling, broiling, frying, steaming, or grilling;

- Mixing;

- Addition of gas such as air entrainment for bread or gasification of soft drinks;

- Proofing;

- Spray drying;

- Pasteurization;

- Packaging.

Performance Parameters for Food Processing

When designing processes for the food industry, the following performance parameters may be taken into account:

- Hygiene, measured, for instance, by the number of microorganisms per ml of finished product;

- Energy consumption, measured, for instance, by "ton of steam per ton of sugar produced";

- Minimization of waste, measured, for instance, by the "percentage of peeling loss during the peeling of potatoes";

- Labor used, measured, for instance, by the "number of working hours per ton of finished product";

- Minimization of cleaning stops, measured, for instance, by the "number of hours between cleaning stops".

Benefits

More and more people live in the cities far away from where food is grown and produced. In many families, the adults are work from home and therefore there is little time for the preparation of food based on fresh ingredients. The food industry offers products that fulfill many different needs: from peeled potatoes that simply need to be boiled at home to fully prepared ready meals that can be heated up in the microwave oven in a few minutes.

Benefits of food processing include toxin removal, preservation, easing marketing and distribution tasks, and increasing food consistency. In addition, it increases seasonal availability of many foods, enables transportation of delicate perishable foods across long distances, and makes many kinds of foods safe to eat by de-activating spoilage and pathogenic micro-organisms. Modern supermarkets would not be feasible without modern food processing techniques, long voyages would not be possible, and military campaigns would be significantly more difficult and costly to execute.

Microwave oven

Modern food processing also improves the quality of life for allergy sufferers, diabetics, and other people who cannot consume some common food elements. Food processing can also add extra nutrients such as vitamins.

Processed foods are often less susceptible to early spoilage than fresh foods, and are better suited for long distance transportation from the source to the consumer. Fresh materials, such as fresh produce and raw meats, are more likely to harbor pathogenic microorganisms (for example, Salmonella) capable of causing serious illnesses.

Drawbacks

In general, fresh food that has not been processed other than by washing and simple kitchen preparation, may be expected to contain a higher proportion of naturally occurring vitamins, fiber and minerals than the equivalent product processed by the food industry. Vitamin C, for example, is destroyed by heat and therefore canned fruits have a lower content of vitamin C than fresh ones.

Tate & Lyle brand Corn Syrup being moved by tank car.

Food processing can lower the nutritional value of foods. Processed foods tend to include food additives, such as flavorings and texture enhancing agents, which may have little or no nutritive value, and some may be unhealthy. Some preservatives added or created during processing, such as nitrites or sulfites, may cause adverse health effects.

Processed foods often have a higher ratio of calories to other essential nutrients than unprocessed foods, a phenomenon referred to as "empty calories." Most junk foods are processed, and fit this category.

High quality and hygiene standards must be maintained to ensure consumer safety, and failure to maintain adequate standards can have serious health consequences.

Processing food is a very costly process, thus increasing the prices of foods products.

Trends in Modern Food Processing

Health

- Reduction of fat content in final product, for example, by using baking instead of deep-frying in the production of potato chips.

- Maintaining the natural taste of the product, for example, by using less artificial sweetener.

Hygiene

The rigorous application of industry and government endorsed standards to minimize possible risk and hazards. In the U.S., the standard adopted is HACCP.

Efficiency

- Rising energy costs lead to increasing usage of energy-saving technologies, for example, frequency converters on electrical drives, heat insulation of factory buildings, and heated vessels, energy recovery systems.

- Factory automation systems (often Distributed control systems) reduce personnel costs and may lead to more stable production results.

Industries

Food processing industries and practices include the following:

- Cannery
- Industrial rendering
- Meat packing plant

- Slaughterhouse
- Sugar industry
- Vegetable packing plant

Mincing

Mincing food means cutting into pieces about 1/8" or 1/16" in diameter. This is the smallest size without cutting food into a pulp or puree. Use a very sharp knife when mincing. You can cut food into pieces about 1/2", then, using a chef's knife, run the knife over the pieces, occasionally stopping to move the food around with your fingers, until the size is uniform.

Most foods that are minced are used for seasoning, such as onions, garlic, and celery. You can also mince foods in a food processor or blender, but it can be difficult to stop before the food is pureed.

Some food is minced because it will be evenly distributed in a recipe, such as onions in a stew. Other foods are minced because their flavor will be stronger, such as garlic. Whole cloves or sliced garlic will be milder than garlic that is crushed or minced.

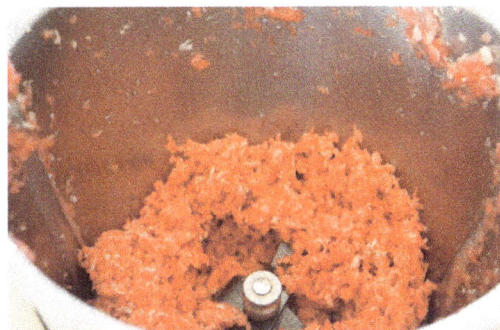

The Advantages of Mincing

The total yield of flesh of low bone content is higher than with filleting alone; up to twice as much can be recovered by separating flesh directly from headless gutted fish. When the fish are first filleted, an additional 8-12 per cent flesh can be separated from the filleting waste.

Some people do not like fatty fish such as herring and mackerel partly because of the large numbers of small bones remaining in the fillets; mince made from these fish is relatively free from bones and might therefore be more widely acceptable. Flesh from underexploited species such as blue whiting, which are difficult to fillet economically because of their small size or awkward shape, can readily be removed in a bone separator.

Mincing offers an opportunity to exercise greater control over flavour, appearance and keeping quality by the incorporation of additives. Rancidity in fatty fish, for example, can be controlled more easily in minced flesh by intimate mixing with permitted antioxidants, or minces of different fat content can be mixed together to give a more desirable result. Mince can be moulded into different shapes, and lends itself to continuous production methods.

The Disadvantages of Mincing

When fish flesh is minced the texture, flavour and sometimes colour are also changed; hence minced fish, and the products derived from it, have at present only limited outlets. Small amounts are used in fish cakes and in less expensive grades of fish finger, and some is used to fill voids in frozen laminated blocks of fillets from which portions and fingers are cut. The present market for mince is small compared with the amount of mince that could be made available from all suitable species.

Mince spoils faster than fillets of the same material, mainly because the structure of the flesh is destroyed during separation, and extra care has to be taken to maintain good quality; in particular, the fish used for making mince has to be initially of very high quality, and processing has to be completed quickly.

Preparation of Fish for Mincing

Whole white fish such as cod and haddock should be gutted and headed. In addition the section of backbone immediately above the belly cavity should be removed; otherwise blood along the backbone causes discoloration and spoilage of the mince, and large pieces of bone tend to damage the rubber belt of the separator. Fillet trimmings can be fed directly into the separator. Whole fatty fish should be gutted and headed, and then split to make them easier to pass through the separator. The skin should be removed from soft-skinned species before putting them through the separator; otherwise too many fragments of skin will pass through with the mince.

AH fish or trimmings being prepared for separation should be washed to remove adhering debris and drained before being fed into the machine.

Mixed species should not be put through the separator unless it is known that there is no adverse interaction; otherwise there is a risk that enzyme activity may result in chemical changes that give an unacceptable texture.

Operation of the Separator

The separator should be cleaned and cooled before use and at intervals during use; feeding flake ice through the machine while running, followed by washing with a high pressure hose, is a simple and effective cleaning and cooling procedure. If the machine becomes clogged during operation, it should be stripped down, a fairly simple task, and the parts washed thoroughly with a hose.

Regular lubrication is essential; the machines are fitted with grease nipples, and a suitable edible grease should be injected through these every few hours when the machine is in daily use.

Fish can be fed from the hopper into the separator in a random manner, but performance can be improved by putting fish in with the skin side against the belt and the cut surface of the flesh against the drums; skins are removed cleanly, the separator is less likely to clog, and the mince is less likely to contain scales and pieces of skin.

Keeping Quality of Mince

Mince made in a hygienic manner has the same initial quality as the raw material from which it was made; when mince is made from fresh white fish, fillets yield the best mince, followed by fillet trimmings, frames or skeletons from which the fillets have been cut, and backbones only, in that descending order of quality. Mince made from stale fish is poor, no matter what parts of the fish are used.

Mince made from fillets cut from 4-day-old iced cod has an unacceptably high bacterial count after 24 hours storage at 5-10°C. Mince made from fillets cut from 12-day-old cod has a high initial count which changes little during 24 hours storage. Hence mince should be made only from fresh raw material, handled quickly and hygienically, and then frozen.

Handling and Storing Mince

Mince should be frozen as soon as it is made, or incorporated in products and then frozen within 4-5 hours of manufacture; products should be kept chilled while awaiting freezing.

Mince can be frozen in blocks in cartons in a horizontal plate freezer in the same manner as for laminated fillet blocks.

The storage life of frozen mince made from good quality cod and haddock flesh is at least 6 months at - 30°C, or 3 months at - 20°C, without any significant loss of quality, but some fish, notably hake and Alaska pollack, have been found to have a shorter cold store life. Minces of all species that include a proportion of active tissue from the backbone and gut will also have a reduced cold storage life. Mince washed with water keeps better than unwashed mince, and mince incorporating additives such as sucrose or sorbitol keeps better than mince without additives. Minces made from fatty fish require protection against oxidation in cold storage.

Sensory Properties of Mince

The appearance and texture of mince are different from those of fillets because the flesh is fragmented, but limited tests have shown that the consumer is not unduly deterred by the unfamiliar form of mince when presented in fish fingers provided it is made from fresh raw material; indeed in tasting tests children showed an apparent preference for fish fingers made from mince rather than fillet.

There is some loss of the sweet flavour of fresh fish during mincing and, more seriously, a slight 'cardboardy' flavour can sometimes be detected, which is more usually associated with the cold storage of whole fish. There may also be a slight increase in firmness and dryness.

Species	Part offish	Weight of bone mg/kg of mince
Cod	Fillets	84
	Trimmings	272
	Frames	1736
	Backbones	4050
Herring	Split fish	2610

When blood-rich tissue from beneath the backbone is included among the raw material, the resulting mince will be red; the mince turns brown when cooked, and has an unacceptable and sometimes extremely objectionable metallic flavour.

Bone separators do not remove all pieces of bone; the table shows typical average bone contents of minces made from different parts of fish, using a 5 mm drum.

Some of the residual bones in mince from fillets or gutted whole fish are needle shaped and sometimes more than 6 mm long, which exceeds some specifications for limits of bone. Bone particles in mince made from frames, that is the skeletons of whole fish from which fillets have been removed, but which still carry some flesh, are blunt and irregular in shape; these would meet a specification that bones should not be capable of piercing the soft palate, but might not meet a specification limiting the permitted weight of bone present. Use of a drum with smaller perforations reduces bone content, but also yields a mince of poorer texture; perforations from 1 to 7 mm are available commercially, but a 5 mm drum generally offers the most reasonable compromise.

Utilization of Mince

Existing uses are in the manufacture of fish cakes, as a filler in laminated fillet blocks with up to 15 per cent mince, and for the manufacture of cheaper fish fingers and portions.

It is possible to make highly acceptable cook-freeze dishes from mince. Fatty fish mince can be used for a range of flavoured fish finger products by adding a suitable sauce, for example tomato, curry, mustard or cheese. Mince is also suitable for making speciality products like Gefüllte Fish. Mince made experimentally from blue whiting has been found suitable for making surimi and kamaboko, two important Japanese fish products; a valuable UK export trade could possibly emerge to meet this market.

Good quality frozen mince in small blocks could find a retail market for use from home freezers as an ingredient of homemade dishes.

Mince lends itself to considerable modification of texture, flavour and appearance, and to the addition of stabilizers and additives; given imaginative development and marketing there may be good prospects of introducing a wide range of commercial food products from minced fish.

Liquefaction

Liquefaction is a common method in food processing used to convert any solid food into a liquid state. Juicing of fruits is one such example.

Liquidiser

A blender or liquidiser is a kitchen and laboratory appliance used to mix, puree, or emulsify food and other substances. A stationary blender consists of a blender jar with a rotating metal blade at the bottom, powered by an electric motor in the base. Some powerful models can also crush ice. The newer immersion blender configuration has a motor on top connected by a shaft to a rotating blade at the bottom, which can be used with any container.

Macerating

Maceration is a process of breaking down and softening various substances. In food preparation, the process most often occurs when soaking fruit in sugar, alcohol, or other flavored liquids with the goals of softening and flavoring the fruit.

Maceration changes a fruit's flavor and texture and is useful for improving the texture of hard, underripe fresh fruit as well as for flavoring fruit at the peak of ripeness. When fruit is macerated, it softens and releases some of its flavor and aroma compounds through its plasmodesmata, which are small channels in the wall of a plant cell that allow molecules and substances to pass through (both in and out) as needed. After maceration, fruit becomes something new, a complex mingling of flavors and textures. The soft fruit and liquid have many uses: a tasty dessert on its own topped with a dollop of whipped cream; a sauce for ice cream or cake; or a filling for pie or a cake where it adds not only flavor, but color and moisture.

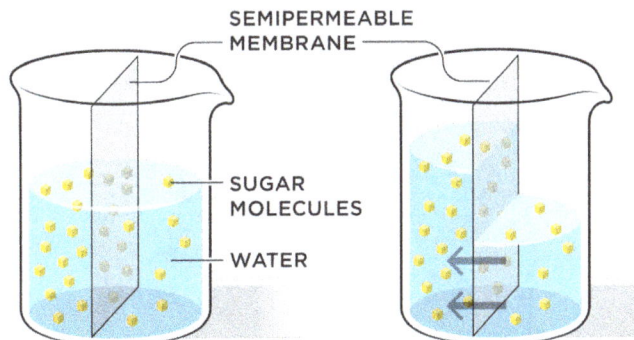

Osmosis is a process in which a fluid flows through a semipermeable membrane (such as a cell wall) from an area of lower concentration to one of higher concentration. Salt and sugar are two of the most common catalysts for osmosis, so putting either ingredient into your maceration mixture will trigger the process. When either comes in contact with food, it works to reach a state of equilibrium with the water content of the food itself to balance the concentration of the salt or sugar in the solution, so beginning the process of osmosis; available water contained within the cells of the fruit is drawn out. The loss of water from the fruit causes it to soften; it also concentrates its natural flavors.

Osmosis works pretty much the same way when fruit is macerated in alcohol because alcohol, like sugar, is hydrophilic and will "steal" water from the fruit, with other flavor molecules along for the ride.

Sugar is hygroscopic, meaning that it attracts and bonds with water. When you macerate with sugar, the water in the fruit is drawn out into the surrounding sugar. As water leaves the fruit, its cells lose volume, reducing the internal pressure on the fruit's cell walls, which then relax, causing the fruit to soften.

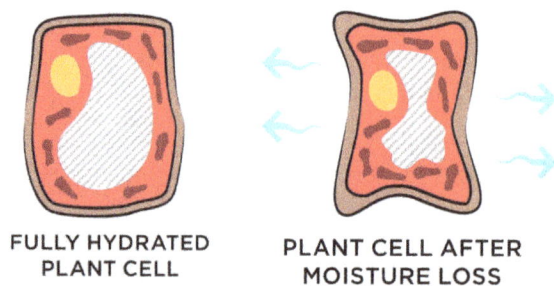

FULLY HYDRATED
PLANT CELL

PLANT CELL AFTER
MOISTURE LOSS

Liquids for maceration include liquors, liqueurs, wine, fruit juice, vinegars, and water that may (and in the case of water must) be infused with all sorts of flavorings, like spices, herbs, tea, and coffee. When it's sprinkled on fruit, sugar draws out moisture, creating its own juicy medium for maceration. Liquids for maceration are sometimes heated to hasten the process or to infuse flavors prior to maceration but maceration does not require the application of heat.

Liquids that have already been infused with flavor provide the fastest flavor transfer to the macerating fruit. For instance, it's best to macerate fruit in a cinnamon-infused liquid rather than adding a cinnamon stick directly to a cold water-based maceration liquid. However, alcohol extracts flavor really effectively. When using a liquid with a high alcohol content, such as whiskey, you can add any mix of whole spices, aromatics, or other flavorful solids because their flavors will be extracted without heating.

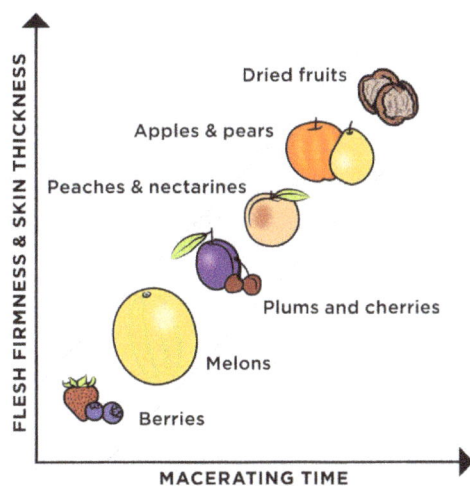

The length of time for macerating fruit can vary from about 30 minutes to a couple of days. To calculate an approximate time, consider a few factors: the thickness of the fruit's skin, the softness of the flesh, and the resulting softness you want for the dish you're preparing. If you are macerating several fruits together, consider the order in which you add the fruit to the liquid. Start firmer-fleshed or tougher-skinned (if unpeeled) fruits first to give them a head start.

Emulsification

Emulsifying is the process step of mixing two or more normally unmixable or unblendable liquid phases in an overall product system or formulation. Emulsification is widely used in industry applications involving chemical processing, liquid pharmaceutical products and food & beverage processing. While a range of emulsifying mixer designs can be used to produce emulsions, such as stirred vessels/agitators, static mixers and homogenizers, rotor-stator mixers are particularly well suited for emulsifying (especially in the macro-emulsion droplet size range).

Often, emulsions do not persist in the emulsified stage for a satisfactory length of time — they revert into the original phases comprising the emulsion. For an emulsion to remain stable, sufficient energy input is required. Generally, the finer the droplets produced, the more stable the emulsion.

High shear mixing plays a controlling role in emulsification by breaking down droplets while simultaneously limiting the re-combining of droplets during processing. It also improves the suspension characteristics of the product.

If an existing emulsion requires increased stability, shearing the emulsion further decreases the droplet size even more; this process of homogenization results in smaller and more uniform droplet size. However, this can lead to subsequent processing using a high-pressure homogenizer, which adds to the cost of production.

Appearance and Properties

Emulsions contain both a dispersed and a continuous phase, with the boundary between the phases called the "interface". Emulsions tend to have a cloudy appearance because the many phase interfaces scatter light as it passes through the emulsion. Emulsions appear white when all light is scattered equally. If the emulsion is dilute enough, higher-frequency (low-wavelength) light will be scattered more, and the emulsion will appear bluer – this is called the "Tyndall effect". If the emulsion is concentrated enough, the color will be distorted toward comparatively longer wavelengths, and will appear more yellow. This phenomenon is easily observable when comparing skimmed milk, which contains little fat, to cream, which contains a much higher concentration of milk fat. One example would be a mixture of water and oil.

Two special classes of emulsions – microemulsions and nanoemulsions, with droplet sizes below 100 nm – appear translucent. This property is due to the fact that light waves are scattered by the droplets only if their sizes exceed about one-quarter of the wavelength of the incident light. Since the visible spectrum of light is composed of wavelengths between 390 and 750 nanometers (nm), if the droplet sizes in the emulsion are below about 100 nm, the light can penetrate through the emulsion without being scattered. Due to their similarity in appearance, translucent nanoemulsions and microemulsions are frequently confused. Unlike translucent nanoemulsions, which require specialized equipment to be produced, microemulsions are spontaneously formed by "solubilizing" oil molecules with a mixture of surfactants, co-surfactants, and co-solvents. The required surfactant concentration in a microemulsion is, however, several times higher than that in a translucent nanoemulsion, and significantly exceeds the concentration of the dispersed phase. Because of many undesirable side-effects caused by surfactants, their presence is disadvantageous or prohibitive in many applications. In addition, the stability of a microemulsion is often easily compromised by dilution, by heating, or by changing pH levels.

Common emulsions are inherently unstable and, thus, do not tend to form spontaneously. Energy input – through shaking, stirring, homogenizing, or exposure to power ultrasound – is needed to form an emulsion. Over time, emulsions tend to revert to the stable state of the phases comprising the emulsion. An example of this is seen in the separation of the oil and vinegar components of vinaigrette, an unstable emulsion that will quickly separate unless shaken almost continuously. There are important exceptions to this rule – microemulsions are thermodynamically stable, while translucent nanoemulsions are kinetically stable.

Whether an emulsion of oil and water turns into a "water-in-oil" emulsion or an "oil-in-water" emulsion depends on the volume fraction of both phases and the type of emulsifier present. In general, the Bancroft rule applies. Emulsifiers and emulsifying particles tend to promote dispersion of the phase in which they do not dissolve very well. For example, proteins dissolve better in water than in oil, and so tend to form oil-in-water emulsions (that is, they promote the dispersion of oil droplets throughout a continuous phase of water).

The geometric structure of an emulsion mixture of two lyophobic liquids with a large concentration of the secondary component is fractal: Emulsion particles unavoidably form dynamic inhomogeneous structures on small length scale. The geometry of these structures is fractal. The size of elementary irregularities is governed by a universal function which depends on the volume content of the components.

Instability

Emulsion stability refers to the ability of an emulsion to resist change in its properties over time. There are four types of instability in emulsions: flocculation, creaming, coalescence, and Ostwald ripening. Flocculation occurs when there is an attractive force between the droplets, so they form flocs, like bunches of grapes. Coalescence occurs when droplets bump into each other and combine to form a larger droplet, so the average droplet size increases over time. Emulsions can also undergo creaming, where the droplets rise to the top of the emulsion under the influence of buoyancy, or under the influence of the centripetal force induced when a centrifuge is used. Creaming is a common phenomenon in dairy and non-dairy beverages (i.e. milk, coffee milk, almond milk, soy milk) and usually does not change the droplet size.

An appropriate "surface active agent" (or "surfactant") can increase the kinetic stability of an emulsion so that the size of the droplets does not change significantly with time. It is then said to be stable. For example, oil-in-water emulsions containing mono- and diglycerides and milk protein as surfactant showed that stable oil droplet size over 28 days storage at 25°C.

Monitoring Physical Stability

The stability of emulsions can be characterized using techniques such as light scattering, focused beam reflectance measurement, centrifugation, and rheology. Each method has advantages and disadvantages.

Accelerating Methods for Shelf Life Prediction

The kinetic process of destabilization can be rather long – up to several months, or even years for some products. Often the formulator must accelerate this process in order to test products in a reasonable time during product design. Thermal methods are the most commonly used – these consist of increasing the emulsion temperature to accelerate destabilization (if below critical temperatures for phase inversion or chemical degradation). Temperature affects not only the viscosity but also the inter-facial tension in the case of non-ionic surfactants or, on a broader scope, interactions of forces inside the system. Storing an emulsion at high temperatures enables the simulation of realistic conditions for a product (e.g., a tube of sunscreen emulsion in a car in the summer heat), but also to accelerate destabilization processes up to 200 times.

Mechanical methods of acceleration, including vibration, centrifugation, and agitation, can also be used.

These methods are almost always empirical, without a sound scientific basis.

Emulsifiers

An emulsifier (also known as an "emulgent") is a substance that stabilizes an emulsion by increasing its kinetic stability. One class of emulsifiers is known as "surface active agents", or surfactants. Emulsifiers are compounds that typically have a polar or hydrophilic (i.e. water-soluble) part and a non-polar (i.e. hydrophobic or lipophilic) part. Because of this, emulsifiers tend to have more or less solubility either in water or in oil. Emulsifiers that are more soluble in water (and conversely, less soluble in oil) will generally form oil-in-water emulsions, while emulsifiers that are more soluble in oil will form water-in-oil emulsions.

Examples of food emulsifiers are:

- Egg yolk – in which the main emulsifying agent is lecithin. In fact, *lecithos* is the Greek word for egg yolk.

- Mustard – where a variety of chemicals in the mucilage surrounding the seed hull act as emulsifiers.

- Soy lecithin - is another emulsifier and thickener.

- Pickering stabilization – uses particles under certain circumstances.

- Sodium phosphates.

- Mono- and diglycerides - a common emulsifier found in many food products (coffee creamers, ice-creams, spreads, breads, cakes).

- Sodium stearoyl lactylate.

- DATEM (Diacetyl Tartaric (Acid) Ester of Monoglyceride) – an emulsifier used primarily in baking.

Detergents are another class of surfactant, and will interact physically with both oil and water, thus stabilizing the interface between the oil and water droplets in suspension. This principle is exploited in soap, to remove grease for the purpose of cleaning. Many different emulsifiers are used in pharmacy to prepare emulsions such as creams and lotions. Common examples include emulsifying wax, polysorbate 20, and ceteareth 20.

Sometimes the inner phase itself can act as an emulsifier, and the result is a nanoemulsion, where the inner state disperses into "nano-size" droplets within the outer phase. A well-known example of this phenomenon, the "Ouzo effect", happens when water is poured into a strong alcoholic anise-based beverage, such as ouzo, pastis, absinthe, arak, or raki. The anisolic compounds, which are soluble in ethanol, then form nano-size droplets and emulsify within the water. The resulting color of the drink is opaque and milky white.

Food emulsifiers act as an interface between the conflicting components of food like water and oil. While preparing the food, often conflicting natural components of food have to be combined into a consistent and pleasing blend. Each component of food (carbohydrate, protein, oil and fat, water, air, etc.) has its own properties which are sometimes conflicting to one another just like oil and water. To make the two components compatible, emulsifiers are used.

Types of Food Emulsifiers

The most frequently used raw materials for emulsifiers include palm oil, rapeseed oil, soy bean oil, sunflower oil or lard/tallow. Egg happens to be the oldest emulsifier. Basic emulsifier production involves combining oil (triglyceride) with glycerol that results in monoglyceride. The type of triglyceride used in the reaction determines the type of emulsifier obtained. Unsaturated triglycerides produce fluid products such as oil while saturated triglycerides result in pasty or solid structures like butter. Monoglycerides can be combined with other substances, such as citric acid and lactic acid, in order to increase their emulsifying properties. Food drugs and cosmetics and pigment emulsions also require one or other kind of emulsifier.

On the basis of their hydrophilic groups, there are basically four categories of natural food emulsifiers and emulsifiers. These are

- Anionics
- Non-ionics
- Cationics
- Amphoterics

Food Emulsifier

- Food Preservatives
- Food Preservatives
- Egg Yolk emulsifying agent lecithin
- Honey
- Mustard
- Soy lecithin
- CSL Calcium Stearoyl Di Laciate
- PolyGlycerol Ester (PGE)
- Sorbitan Ester (SOE)
- PG Ester (PGME)
- Sugar Ester (SE)
- Monoglyceride (MG)
- Acetylated Monoglyceride (AMG)
- Lactylated Monoglyceride (LMG)

Applications of Food Emulsifiers

Food emulsifiers make the food very appealing as without emulsifier the water and the oil content in food will look separate, which will give very unappealing appearance. Apart from this they impart the freshness and quality to the food. Natural food emulsifiers also prevent the growth of moulds in food.

Emulsifiers are used in creams and sauces, bakery, and dairy products. They may be derived from the natural products or chemicals. Common emulsifiers are lecithins, mono- and di-glycerides of fatty acids esters of monoglycerides of fatty acids and phosphated monoglycerides.

Natural food emulsifiers are used in variety of foods. Some basic foods having food emulsifiers are:

- Biscuits
- Extruded snacks
- Cakes
- Soft Drinks
- Toffees
- Frozen Desserts
- Bread

- Margarine

- Coffee Whitener

- Caramels

Preservation

The term food preservation refers to any one of a number of techniques used to prevent food from spoiling. It includes methods such as canning, pickling, drying and freeze-drying, irradiation, pasteurization, smoking, and the addition of chemical additives. Food preservation has become an increasingly important component of the food industry as fewer people eat foods produced on their own lands, and as consumers expect to be able to purchase and consume foods that are out of season.

The vast majority of instances of food spoilage can be attributed to one of two major causes:

1) The attack by pathogens (disease-causing microorganisms) such as bacteria and molds.

2) Oxidation that causes the destruction of essential biochemical compounds and/or the destruction of plant and animal cells. The various methods that have been devised for preserving foods are all designed to reduce or eliminate one or the other (or both) of these causative agents.

For example, a simple and common method of preserving food is by heating it to some minimum temperature. This process prevents or retards spoilage because high temperatures kill or inactivate most kinds of pathogens. The addition of compounds known as BHA and BHT to foods also prevents spoilage in another different way. These compounds are known to act as antioxidants, preventing chemical reactions that cause the oxidation of food that results in its spoilage. Almost all techniques of preservation are designed to extend the life of food by acting in one of these two ways.

The search for methods of food preservation probably can be traced to the dawn of human civilization. People who lived through harsh winters found it necessary to find some means of insuring a food supply during seasons when no fresh fruits and vegetables were available. Evidence for the use of dehydration (drying) as a method of food preservation, for example, goes back at least 5,000 years. Among the most primitive forms of food preservation that are still in use today are such methods as smoking, drying, salting, freezing, and fermenting.

Early humans probably discovered by accident that certain foods exposed to smoke seem to last longer than those that are not. Meats, fish, fowl, and cheese were among such foods. It appears that compounds present in wood smoke have antimicrobial actions that prevent the growth of organisms that cause spoilage. Today, the process of smoking has become a sophisticated method of food preservation with both hot and cold forms in use. Hot smoking is used primarily with fresh or frozen foods, while cold smoking is used most often with salted products. The most advantageous conditions for each kind of smoking—air velocity, relative humidity, length of exposure, and salt content, for example—are now generally understood and applied during the smoking process. For example, electrostatic precipitators can be employed to attract smoke particles and improve the penetration of the particles into meat or fish. So many alternative forms of preservation are now available that smoking no longer holds the position of importance it once did with ancient peoples. More frequently, the process is used to add interesting and distinctive flavors to foods.

Because most disease-causing organisms require a moist environment in which to survive and multiply, drying is a natural technique for preventing spoilage. Indeed, the act of simply leaving foods out in the sun and wind to dry out is probably one of the earliest forms of food preservation. Evidence for the drying of meats, fish, fruits, and vegetables go back to the earliest recorded human history. At some point, humans also learned that the drying process could be hastened and improved by various mechanical techniques. For example, the Arabs learned early on that apricots could be preserved almost indefinitely by macerating them, boiling them, and then leaving them to dry on broad sheets. The product of this technique, quamaradeen, is still made by the same process in modern Muslim countries.

Today, a host of dehydrating techniques are known and used. The specific technique adopted depends on the properties of the food being preserved. For example, a traditional method for preserving rice is to allow it to dry naturally in the fields or on drying racks in barns for about two weeks. After this period of time, the native rice is threshed and then dried again by allowing it to sit on straw mats in the sun for about three days. Modern drying techniques make use of fans and heaters in controlled environments. Such methods avoid the uncertainties that arise from leaving crops in the field to dry under natural conditions. Controlled temperature air drying is especially popular for the preservation of grains such as maize, barley, and bulgur.

Vacuum drying is a form of preservation in which a food is placed in a large container from which air is removed. Water vapor pressure within the food is greater than that outside of it, and water evaporates more quickly from the food than in a normal atmosphere. Vacuum drying is biologically desirable since some enzymes that cause oxidation of foods become active during normal air drying. These enzymes do not appear to be as active under vacuum drying conditions, however. Two of the special advantages of vacuum drying are that the process is more efficient at removing water from a food product, and it takes place more quickly than air drying. In one study, for example, the drying time of a fish fillet was reduced from about 16 hours by air drying to six hours as a result of vacuum drying.

Coffee drinkers are familiar with the process of dehydration known as spray drying. In this process, a concentrated solution of coffee in water is sprayed though a disk with many small holes in it. The surface area of the original coffee grounds is increased many times, making dehydration of the dry product much more efficient. Freeze-drying is a method of preservation that makes use of the physical principle known as sublimation. Sublimation is the process by which a solid passes directly to the gaseous phase without first melting. Freeze-drying is a desirable way of preserving food because at low temperatures (commonly around 14°F to −13°F [−10°C to −25°C]) chemical reactions take place very slowly and pathogens have difficulty surviving. The food to be preserved by this method is first frozen and then placed into a vacuum chamber. Water in the food first freezes and then sublimes, leaving a moisture content in the final product of as low as 0.5%.

The precise mechanism by which salting preserves food is not entirely understood. It is known that salt binds with water molecules and thus acts as a dehydrating agent in foods. A high level of salinity may also impair the conditions under which pathogens can survive. In any case, the value of adding salt to foods for preservation has been well known for centuries. Sugar appears to have effects similar to those of salt in preventing spoilage of food. The use of either compound (and of certain other natural materials) is known as curing. A desirable side effect of using salt or sugar as a food preservative is, of course, the pleasant flavor each compound adds to the final product.

Curing can be accomplished in a variety of ways. Meats can be submerged in a salt solution known as brine, for example, or the salt can be rubbed on the meat by hand. The injection of salt solutions into meats has also become popular. Food scientists have now learned that a number of factors relating to the food product and to the preservative conditions affect the efficiency of curing. Some of the food factors include the type of food being preserved, the fat content, and the size of treated pieces. Preservative factors include brine temperature and concentration, and the presence of impurities.

Curing is used with certain fruits and vegetables, such as cabbage (in the making of sauerkraut), cucumbers (in the making of pickles), and olives. It is probably most popular, however, in the preservation of meats and fish. Honey-cured hams, bacon, and corned beef ("corn" is a term for a form of salt crystals) are common examples.

Freezing is an effective form of food preservation because the pathogens that cause food spoilage are killed or do not grow very rapidly at reduced temperatures. The process is less effective in food preservation than are thermal techniques such as boiling because pathogens are more likely to be able to survive cold temperatures than hot temperatures. In fact, one of the problems surrounding the use of freezing as a method of food preservation is the danger that pathogens deactivated (but not killed) by the process will once again become active when the frozen food thaws.

A number of factors are involved in the selection of the best approach to the freezing of foods, including the temperature to be used, the rate at which freezing is to take place, and the actual method used to freeze the food. Because of differences in cellular composition, foods actually begin to freeze at different temperatures ranging from about 31°F (−0.6°C) for some kinds of fish to 19°F (−7°C) for some kinds of fruits.

The rate at which food is frozen is also a factor, primarily because of aesthetic reasons. The more slowly food is frozen, the larger the ice crystals that are formed. Large ice crystals have the tendency to cause rupture of cells and the destruction of texture in meats, fish, vegetables, and fruits. In order to deal with this problem, the technique of quick-freezing has been developed. In quick-freezing, a food is cooled to or below its freezing point as quickly as possible. The product thus obtained, when thawed, tends to have a firm, more natural texture than is the case with most slow-frozen foods.

About a half dozen methods for the freezing of foods have been developed. One, described as the plate, or contact, freezing technique, was invented by the American inventor Charles Birdseye in 1929. In this method, food to be frozen is placed on a refrigerated plate and cooled to a temperature less than its freezing point. Alternatively, the food may be placed between two parallel refrigerated plates and frozen. Another technique for freezing foods is by immersion in very cold liquids. At one time, sodium chloride brine solutions were widely used for this purpose. A 10% brine solution, for example, has a freezing point of about 21°F (−6°C), well within the desired freezing range for many foods. More recently, liquid nitrogen has been used for immersion freezing. The temperature of liquid nitrogen is about −320°F (−195.5°C), so that foods immersed in this substance freeze very quickly.

As with most methods of food preservation, freezing works better with some foods than with others. Fish, meat, poultry, and citrus fruit juices (such as frozen orange juice concentrate) are among the foods most commonly preserved by this method.

Fermentation is a naturally occurring chemical reaction by which a natural food is converted into another form by pathogens. It is a process in which food spoils, but results in the formation of an edible product. Perhaps the best example of such a food is cheese. Fresh milk does not remain in edible condition for a very long period of time. Its pH is such that harmful pathogens begin to grow in it very rapidly. Early humans discovered, however, that the spoilage of milk can be controlled in such a way as to produce a new product, cheese.

Bread is another food product made by the process of fermentation. Flour, water, sugar, milk, and other raw materials are mixed together with yeasts and then baked. The addition of yeasts brings about the fermentation of sugars present in the mixture, resulting in the formation of a product that will remain edible much longer than will the original raw materials used in the bread-making process.

Heating food is an effective way of preserving it because the great majority of harmful pathogens are killed at temperatures close to the boiling point of water. In this respect, heating foods is a form of food preservation comparable to that of freezing but much superior to it in its effectiveness. A preliminary step in many other forms of food preservation, especially forms that make use of packaging, is to heat the foods to temperatures sufficiently high to destroy pathogens.

In many cases, foods are actually cooked prior to their being packaged and stored. In other cases, cooking is neither appropriate nor necessary. The most familiar example of the latter situation is pasteurization. During the 1860s, the French bacteriologist Louis Pasteur discovered that pathogens in foods could be destroyed by heating those foods to a certain minimum temperature. The process was particularly appealing for the preservation of milk since preserving milk by boiling is not a practical approach. Conventional methods of pasteurization called for the heating of milk to a temperature between 145 and 149°F (63 and 65°C) for a period of about 30 minutes, and then cooling it to room temperature. In a more recent revision of that process, milk can also be "flash-pasteurized" by raising its temperature to about 160°F (71°C) for a minimum of 15 seconds, with equally successful results. A process known as ultra-high-pasteurization uses even higher temperatures, of the order of 194–266°F (90–130°C), for periods of a second or more.

One of the most common methods for preserving foods today is to enclose them in a sterile container. The term "canning" refers to this method although the specific container can be glass, plastic, or some other material as well as a metal can, from which the procedure originally obtained its name. The basic principle behind canning is that a food is sterilized, usually by heating, and then placed within an air-tight container. In the absence of air, no new pathogens can gain access to the sterilized food. In most canning operations, the food to be packaged is first prepared in some way—cleaned, peeled, sliced, chopped, or treated in some other way—and then placed directly into the container. The container is then placed in hot water or some other environment where its temperature is raised above the boiling point of water for some period of time. This heating process achieves two goals at once. First, it kills the vast majority of pathogens that may be present in the container. Second, it forces out most of the air above the food in the container.

After heating has been completed, the top of the container is sealed. In home canning procedures, one way of sealing the (usually glass) container is to place a layer of melted paraffin directly on top of the food. As the paraffin cools, it forms a tight solid seal on top of the food. Instead of or in addition to the paraffin seal, the container is also sealed with a metal screw top containing a rubber gasket. The first glass jar designed for this type of home canning operation, the Mason jar, was patented in 1858.

The commercial packaging of foods frequently makes use of tin, aluminum, or other kinds of metallic cans. The technology for this kind of canning was first developed in the mid-1800s, when individual workers hand-sealed cans after foods had been cooked within them. At this stage, a single worker could seldom produce more than 100 "canisters" (from which the word "can" later came) of food a day. With the development of far more efficient canning machines in the late nineteenth century, the mass production of canned foods became a reality.

As with home canning, the process of preserving foods in metal cans is simple in concept. The foods are prepared and the empty cans are sterilized. The prepared foods are then added to the sterile metal can, the filled can is heated to a sterilizing temperature, and the cans are then sealed by a machine. Modern machines are capable of moving a minimum of 1,000 cans per minute through the sealing operation.

The majority of food preservation operations used today also employ some kind of chemical additive to reduce spoilage. Of the many dozens of chemical additives available, all are designed either to kill or retard the growth of pathogens or to prevent or retard chemical reactions that result in the oxidation of foods. Some familiar examples of the former class of food additives are sodium benzoate and benzoic acid; calcium, sodium propionate, and propionic acid; calcium, potassium, sodium sorbate, and sorbic acid; and sodium and potassium sulfite. Examples of the latter class of additives include calcium, sodium ascorbate, and ascorbic acid (vitamin C); butylated hydroxyanisole (BHA) and butylated hydroxytoluene (BHT); lecithin; and sodium and potassium sulfite and sulfur dioxide.

A special class of additives that reduce oxidation is known as the sequestrants. Sequestrants are compounds that "capture" metallic ions, such as those of copper, iron, and nickel, and remove them from contact with foods. The removal of these ions helps preserve foods because in their free state they increase the rate at which oxidation of foods takes place. Some examples of sequestrants used as food preservatives are ethylenediamine-tetraacetic acid (EDTA), citric acid, sorbitol, and tartaric acid.

The lethal effects of radiation on pathogens has been known for many years. Since the 1950s, research in the United States has been directed at the use of this technique for preserving certain kinds of food. The radiation used for food preservation is normally gamma radiation from radioactive isotopes or machine-generated x rays or electron beams. One of the first applications of radiation for food preservation was in the treatment of various kinds of herbs and spices, an application approved by the U.S. Food and Drug Administration (FDA) in 1983. In 1985, the FDA extended its approval to the use of radiation for the treatment of pork as a means of destroying the pathogens that cause trichinosis. Experts predict that the ease and efficiency of food preservation by means of radiation will develop considerably in the future. That future is somewhat clouded, however, by fears expressed by some scientists and members of the general public about the dangers that irradiated foods may have for humans. In addition to a generalized concern about the possibilities of being exposed to additional levels of radiation in irradiated foods (not a possibility), critics have raised questions about the creation of new and possibly harmful compounds in food that has been exposed to radiation.

References

- Aulton, Michael E., ed. (2007). Aulton's Pharmaceutics: The Design and Manufacture of Medicines (3rd ed.). Churchill Livingstone. pp. 92–97, 384, 390–405, 566–69, 573–74, 589–96, 609–10, 611. ISBN 978-0-443-10108-3

- Kumar, Harish V.; Woltornist, Steven J.; Adamson, Douglas H. (2016-03-01). "Fractionation and characterization of graphene oxide by oxidation extent through emulsion stabilization". Carbon. 98: 491–495. doi:10.1016/j.carbon.2015.10.083

- The-science-of-maceration: finecooking.com, Retrieved 13 April 2018

- Anne-Marie Faiola (2008-05-21). "Using Emulsifying Wax". TeachSoap.com. TeachSoap.com. Retrieved 2008-07-22

- Loi, Chia Chun; Eyres, Graham T.; Birch, E. John (2019). "Effect of mono- and diglycerides on physical properties and stability of a protein-stabilised oil-in-water emulsion". Journal of Food Engineering. 240: 56–64. doi:10.1016/j.jfoodeng.2018.07.016. ISSN 0260-8774

- Emulsification-process, applications: quadroliquids.com, Retrieved 23 May 2018

- Troy, David A.; Remington, Joseph P.; Beringer, Paul (2006). Remington: The Science and Practice of Pharmacy (21st ed.). Philadelphia: Lippincott Williams & Wilkins. pp. 325–336, 886–87. ISBN 0-7817-4673-6

- Khan, A. Y.; Talegaonkar, S; Iqbal, Z; Ahmed, F. J.; Khar, R. K. (2006). "Multiple emulsions: An overview". Current drug delivery. 3 (4): 429–43. doi:10.2174/156720106778559056. PMID 17076645

- Emulsifiers: foodadditivesworld.com, Retrieved 21 March 2018

- "Nanoemulsion vaccines show increasing promise". Eurekalert! Public News List. University of Michigan Health System. 2008-02-26. Retrieved 2008-07-22

- Mason TG, Wilking JN, Meleson K, Chang CB, Graves SM (2006). "Nanoemulsions: Formation, structure, and physical properties" (PDF). Journal of Physics: Condensed Matter. 18 (41): R635. Bibcode:2006JPCM...18R.635M. doi:10.1088/0953-8984/18/41/R01

- Sports-and-everyday-life, food-and-cooking, food-and-drink, food-preservation-3418500949: encyclopedia.com, Retrieved 18 June 2018

Understanding Food Preservation Techniques

Food preservation is an important process in food production. Its objective is to prevent the growth of microorganism for slowing the oxidation of fats that are responsible for rancidity of food. The topics elaborated in this chapter will help in providing a better perspective about the different food preservation techniques, such as curing, freezing, boiling, pickling, canning, pasteurization, etc. It also elucidates the modern practices such as the use of nonthermal plasma, high-pressure food preservation, biopreservation, pulsed electric field electroporation, etc.

Curing

Curing is the addition to meats of some combination of salt, sugar, nitrite and/or nitrate for the purposes of preservation, flavor and color. Some publications distinguish the use of salt alone as salting, corning or salt curingand reserve the word curing for the use of salt with nitrates/nitrites. The cure ingredients can be rubbed on to the food surface, mixed into foods dry (dry curing), or dissolved in water (brine, wet, or pickle curing). In the latter processes, the food is submerged in the brine until completely covered. With large cuts of meat, brine may also be injected into the muscle. The term pickle in curing has been used to mean any brine solution or a brine cure solution that has sugar added.

Salting/Corning

Salt inhibits microbial growth by plasmolysis. In other words, water is drawn out of the microbial cell by osmosis due to the higher concentration of salt outside the cell. A cell loses water until it reaches a state first where it cannot grow and cannot survive any longer. The concentration of salt outside of a microorganism needed to inhibit growth by plasmolysis depends on the genus and species of the microorganism. The growth of some bacteria is inhibited by salt concentrations as low as 3%, e.g., Salmonella, whereas other types are able to survive in much higher salt concentrations, e.g., up to 20% salt for Staphylococcus or up to 12% salt for Listeria monocytogenes. Fortunately the growth of many undesirable organisms normally found in cured meat and poultry products is inhibited at relatively low concentrations of salt.

Salting can be accomplished by adding salt dry or in brine to meats. Dry salting, also called corning originated in Anglo-Saxon cultures. Meat was dry-cured with coarse "corns" or pellets of salt. Corned beef of Irish fame is made from a beef brisket, although any cut of meat can be corned. Salt brine curing involves the creation of brine containing salt, water and other ingredients such as sugar, erythorbate, or nitrites. Age-old tradition was to add salt to the brine until it floated an egg. Today, however, it is preferred to use a hydrometer or to carefully mix measured ingredients from a reliable recipe. Once mixed and placed into a suitable container, the food is submerged in the salt brine. Brine curing usually produces an end product that is less salty compared to dry curing. Injection of brine into the meat can also speed the curing process.

Nitrate/ Nitrite Curing

Most salt cures do not contain sufficient levels of salt to preserve meats at room temperature and Clostridium botulinum spores can survive. In the early 1800's it was realized that saltpeter ($NaNO_3$ or KNO_3) present in some impure curing salt mixtures would result in pink colored meat rather than the typical gray color attained with a plain salt cure. This nitrate/nitrite in the curing process was found to inhibit growth of Clostridium. Recent evidence indicates that they may also inhibit E. coli, Salmonella, and Campylobacter if in sufficient quantities.

Several published studies indicated that N-nitrosoamines were considered carcinogenic in animals. For this reason, nitrate is prohibited in bacon and the nitrite concentration is limited in other cured meats. In other cured foods, there is insufficient scientific evidence for N-nitrosamine formation and a link to cancer.

Cure Mixtures

For the home food preserver, measuring small batches of cure for nitrites or nitrates would require an analytical scale that few consumers have access to. Therefore, some manufacturers sell premixed salt and nitrate/nitrite curing mixes for easy home use. Caution is needed when using pure saltpeter instead of commercially prepared mixes, since accidental substitution of saltpeter for table salt in recipes can result in lethal toxic levels.

Some examples of commercially prepared cures include:

Prague Powder, Insta Cure or Modern Cure

This cure contains sodium nitrite (6.25%) mixed with salt (93.75%). Consumers are recommended to use 1 oz. for every 25 lb. of meat or one level teaspoon of cure for 5 lb. of meat.

Prague Powder

This mix is used for dry cured meats that require long (weeks to months) cures. It contains 1 oz. of sodium nitrite and 0.64 oz. of sodium nitrate. It is recommended that this cure be combined with each 1 lb. of salt and for products that do not require cooking, smoking, or refrigeration. This cure, which contains sodium nitrate, acts like a time-release cure, slowly breaking down into sodium nitrite, then into nitric oxide. The manufacturer recommends using 1 oz. of cure for 25 lbs. of meat or one level teaspoon of cure for 5 lbs. of meat.

Mixes

Many individual manufacturers and commercial sausage makers produce curing mixtures, often combining sugar and spices with the salt and nitrite/nitrates. It is important that consumers follow manufacturer directions carefully.

Saltpeter, Sodium or Potassium Nitrate

Commercially, nitrate is no longer allowed for use in curing of smoked and cooked meats, non-smoked and cooked meats, or sausages (US FDA 1999). However, nitrate is still allowed in small amounts in the making of dry cured uncooked products. Home food preservers should avoid the direct use of this chemical and opt for the mixtures described.

Combination Curing

Some current recipes for curing have vinegar, citrus juice, or alcohol as ingredients for flavor. Addition of these chemicals in sufficient quantities can contribute to the preservation of the food being cured.

Flavor of Cured Meats

Besides preservation, the process of curing introduces both a desired flavor and color. Cured meat flavor is thought to be a composite result of the flavors of the curing agents and those developed by bacterial and enzymatic action.

Salt

Because of the amount of salt used in most curing processes, the salt flavor is the most predominant.

Sugar

Sugar is a minor part of the composite flavor, with bacon being an exception. Because of the tremendous amount of salt used, sugar serves to reduce the harshness of the salt in cured meat and enhance the sweetness of the product (ie. Sweet Lebanon Bologna). Sugar also serves as a nutrient source for the flavor-producing bacteria of meat during long curing processes.

Spices and Flavor Enhancers

Spices add characteristic flavors to the meats. Recent studies have suggested that some spices can have added preservative effects (Doyle 1999). However, the quantities of spice needed to achieve these effects may be well beyond the reasonable quantities of use.

Nitrates/Nitrites

Nitrites and nitrate conversion to nitrite provide the characteristic cured flavor and color.

Fermentation

The tangy flavor observed in dry fermented sausages, such as pepperoni, is the result of bacterial fermentation or the addition of chemicals such as glucono-δ(*delta*)-lactone.

Smoking

The process of smoking gives the product the characteristic smoky flavor that can be varied slightly with cure recipes and types of smoke used.

Color of Cured Meats

A high concentration of salt promotes the formation of an unattractive gray color within some meat. Nitrate when used for some dry-cured, non-cooked meats is reduced to nitrite then to nitric oxide, which reacts with myoglobin (muscle pigment) to produce the red or pink cured color. If nitrite is used as the curing agent, there is no need for the nitrate reduction step, and the development of the cure color is much more rapid.

Generation of Nitric Oxide (NO):

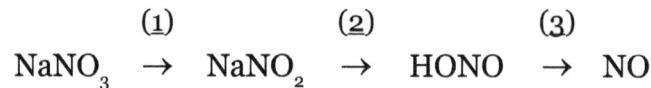

$$\underset{(1)}{NaNO_3} \rightarrow \underset{(2)}{NaNO_2} \rightarrow \underset{(3)}{HONO} \rightarrow NO$$

1. Sodium nitrate is reduced to sodium nitrite by microorganisms such as Micrococcus spp. present on meats.

2. Sodium nitrite is reduced to nitrous acid in the presence of an acidic environment (e.g., by fermentation or by addition of glucono-δ(delta)-lactone).

3. Nitrous acid forms nitric oxide. Nitric oxide reacts with myoglobin (meat pigments) to form a red color.

The time required for a cured color to develop may be shortened with the use of cure accelerators, e.g., ascorbic acid, erythorbic acid, or their derivatives. Cure accelerators tend to speed up chemical conversion of nitric acid to nitric oxide. They also serve as oxygen scavengers, which slow the fading of the cured meat color in the presence of sunlight and oxygen. Some studies have indicated that cure accelerators have antimicrobial properties, especially for the newly emerging pathogens like *E. coli*O157:H7 and *Listeria monocytogenes* (Doyle 1999). Since cure accelerators are rarely used in home curing, this information needs further review or research to determine what benefits home curing would have by using certain cure accelerators.

Ham

Ham is cured pork from the hind leg of the hog. Picnic shoulder or picnic ham is made from the front leg of the hog. Ham varieties may or may not be smoked and are available in many regional and ethnic styles. Curing solutions for hams typically contain salt, sodium nitrate, sugar, and seasonings. Dry-cured ham includes country ham and proscuitto. The dry cure mixture is rubbed onto the pork surface and the meat is cured (at or below 40°F) from weeks to a year or more. During this aging process, the moisture is reduced by 18-25%, making these hams safe at room temperature. Brine-cured ham includes culatello and Irish Hams. Usually the fresh meat is both injected with brine and submerged into the brine to allow the cure to reach all of the meat.

Bacon

Bacon is cured or smoked hog meat from the pig belly. Bacon produced at home, is typically dry-cured with salt, nitrites, sugar, and spices for a week or longer. Because of concern over N-nitrosamines, the use of nitrates for bacon curing is not allowed commercially. Home preparations, such as Morton Smoked-flavored sugar cure, contain nitrates and are recommended by the manufacturer for the use in bacon curing. Some ethnic bacon (Canadian bacon and Irish bacon) is made from leaner cuts. Pancetta is Italian bacon that is not smoked. Salt pork is salted pork belly fat.

Beef

The most well known cured beef product is corned beef made from the beef brisket. Pastrami is smoked corned beef.

Poultry

Any variety of poultry can be cured and smoked. Curing and smoking imparts a unique, delicate flavor and pink color to poultry meat. As with other meats, curing and smoking increases the refrigerated storage life of poultry. When preparing smoked poultry products, most consumers use mild cures (relatively low salt) to maintain the poultry flavor.

Fish

Any fish can be salted and smoked. Some varieties of fish make for better tasting products than others. Commercially, nitrite curing is only allowed for sable, salmon, shad, chub, and tuna in the U.S. Other species were never included in the Code of Federal Regulations simply because industry members did not respond to initial inquiries about GRAS (Generally Regarded as Safe) practices (Ken Hilderbrand, personal communication).

Sausage

Sausage can be made from any meat source, and is typically ground. Sausage can be uncured and unsmoked, but for the purposes of this document, we consider only cured and/or smoked sausage. Usually cure ingredients (salt, nitrates/nitrites, and spices) are mixed with the ground meat and stuffed into casings (animal intestines or collagen). The product is then cured for a short time (e.g. overnight for bologna) at refrigerated temperatures. It may or may not be smoked, dried, or fermented.

Game

Venison, bear, elk, wild boar, wild turkey, rabbit and other game animals can be successfully cured/smoked.

Food Safety Concerns

Concern for food safety has arisen over:

1) The public's desire for variety and healthfulness that leads them to both non-traditional foods and non-traditional processes that may lack research into their safety.

2) The emergence of new foodborne diseases that challenge the safety of traditional food preservation methods. Bacteria, yeasts and molds find meat a suitable substrate for growth, resulting in meat quality and safety deterioration. Foodborne diseases are mostly of bacterial origin and meat has been implicated in roughly one third of the foodborne outbreaks in North America. The pathogenic microorganisms representing the greatest risk with meat

and poultry borne diseases are *Salmonella* spp., *Campylobacter* spp, verotoxigenic *Escherichia coli*, *Listeria monocytogenes* and *Toxoplasma gondii*. Consumers and home food preservers should be warned that microorganisms are ubiquitous in the environment and that pathogens may survive traditional and non-traditional food preservation techniques if they are improperly processed.

Non-traditional Foods and Non-traditional Processes

Today, consumers demand foods that are minimally processed, as "natural" as possible, and yet are convenient to use. Complicating these factors is a consumer preference toward cured and smoked foods that are processed with lower salt, lower nitrate and higher moisture levels. These parameters have a tremendous impact on the safety of a given cured/smoked food or process. Preferences for low fat and low sugar have less impact on the safety, but these factors can change the traditional curing and smoking process. It will be difficult to completely eliminate the use of nitrite, as there is no known substitute for it as a curing agent for meat. Nonetheless, the demand for fewer chemicals added to foods has put pressure on the industry and the scientific community to seek new alternatives.

In-home vacuum packaging machines have become popular in recent years. It is important to realize that in-home vacuum packaging is not a substitution for cooking or any form of food preservation, e.g., refrigeration, freezing, or curing. In-home vacuum packaging can reduce the quality deterioration of foods catalyzed by oxygen, such as rancidity. Many food spoilage and food poisoning organisms require oxygen for growth and would also be inhibited by this process. However, the most deadly food poisoning organism, *Clostridium botulinum* requires a low oxygen atmosphere and therefore, vacuum packaging favors its growth. In cured meats, careful attention must be paid to proper use of nitrates/nitrites that inhibit *Clostridium botulinum* prior to use of in-home vacuum packagers. To further reduce the risk of botulism after vacuum packaging, properly refrigerate the cured/smoked meats. Under normal processing, freezing of salt-cured meats is not recommended, due to oxidative rancidity that affects the quality and flavor of the product.

Emergence of New Foodborne Diseases

More than 200 known diseases are transmitted through food .The causes include viruses, bacteria, and parasites. Many of the pathogens causing foodborne illness were not recognized 20 years ago. Major emerging pathogens include *Campylobacter jejuni*, *Salmonella*, *Listeria monocytogenes*, and *Escherichia coli* O157:H7. Many emerging foodborne diseases can cause chronic and serious health problems.

Food Poisoning Organisms

Microorganisms are ubiquitous in foods. Some can be present and harmless. Others can be present and produce chemicals that alter the acceptability of the food, hence food spoilage. Lastly, microorganisms can be present where they themselves or the products they produce can cause food poisoning.

Botulism

The majority (65%) of botulism cases are a result of inadequate home food processing or preservation. Botulism results from ingestion of a toxin produced by the bacterium *C. botulinum*. This bacterium requires a moist, oxygen-free environment, low acidity (pH greater than 4.6) and temperatures in the danger zone (38-140°F) to grow and produce toxin. *C. botulinum* forms heat resistant spores that can become dangerous if allowed to germinate, grow, and produce toxin. Sufficient heat can be used to inactivate the toxin (180°F for 4 min., Kendall 1999). *C. botulinum* thrives in moist foods that are low in salt (less than 10%), particularly when they are stored at temperatures above 38°F. These organisms will not grow in an aerobic environment, but other aerobic organisms in a closed system can rapidly convert an aerobic environment to an anaerobic environment by using the oxygen for their own growth, permitting growth of *C. botulinum*.

Clostridium Perfringens

Spores of some strains of *Clostridium perfringens* are so heat resistant that they survive boiling for four or more hours. Furthermore, cooking drives off oxygen, kills competitive organisms, and heat-shocks the spores, all of which promote germination to vegetative or growing cells. Once the spores have germinated, a warm, moist, protein-rich environment with little or no oxygen is necessary for growth. If such conditions exist (i.e., incorrectly holding meats at warm room temperature for smoking), sufficient numbers of vegetative cells may be produced to cause illness upon ingestion of the contaminated meat product.

Listeria Monocytogenes

L. monocytogenes has been found in fermented raw-meat sausages, raw and cooked poultry, raw meats (all types), and raw and smoked fish. Its ability to grow at temperatures as low as 3°C, permits multiplication in refrigerated foods. The organism grows in the pH range of 5.0 to 9.5 and is resistant to freezing. It is salt tolerant and relatively resistant to drying, but easily destroyed by heat. (It grows between 34 - 113°F).

E. coli O157:H7

Ground beef is the food most associated with *E. coli* O157:H7 outbreaks, but smoked and cured foods also have been implicated, including dry-cured salami, game meat, and homemade venison jerky. Studies have shown that *E. coli* O157:H7 can survive the typical dry fermentation processing conditions; *E. coli*O157:H7's tolerance of acidic conditions has also been reported in the processing of other foods such as apple cider and mayonnaise. These findings led to significant changes in the food industry and in the manufacturing of dry fermented sausage in the U.S. In August 1995, USDA/FSIS recommended using a heat process (145°F for 4 minutes) to inhibit *E. coli* O157:H7 growth in sausage.

Trichinosis

Details on trichinosis can be found in a publication by the National Pork Producers Council (Gamble) and on trichinosis statistics in the USA (CDC 1988). Trichinosis is an infestation of trichinae, or *Trichinella spiralis* or other *Trichinella* spp. The parasites invade the muscles causing severe pain and edema. It can be avoided by ensuring that cooked pork or certain wild game meat reaches an internal temperature of 150°F or more. Freezing the pork according to the following chart also can kill trichinae:

Freezer Temperature	Group 1 Days	Group 2 Days
5°F	20	30
-10°F	10	20
-20°F	6	12
Group 1 comprises product in separate pieces not exceeding 6" in thickness or arranged on separate racks with the layers not exceeding 6" in depth. Group 2 comprises product in pieces, layers or within containers the thickness of which exceeds 6" but not 27".		

Freezing Pork to Kill Trichinae

Although the incidence of trichinosis has decreased markedly from 300 to 400 cases annually in the 1940's to less than 90 cases per year in the early 1980's, this disease remains a problem in the United States. According to USDA recommendations, *T. spiralis* in pork is rendered non-viable if held at 5°F, a temperature achievable in noncommercial freezers, for 20 days. However, meat from wild game, such as polar bear or walrus meat that has been infected with *T. spiralis*, remains infective even after 24 months of storage at 0°F. The difference in susceptibility may be caused by different strains of *T. spiralis* found in domestic versus wild animals. Adequate cooking (170°F. internally), well above the thermal death point of the organism (137°F), remains the best safeguard against trichinosis in game meats.

Staphylococcus Aureus

Staphylococcus is more salt-tolerant than most other bacteria. It is naturally present on human skin. Some species of *Staphylococcus* produce toxins that cause food poisoning. So, handling of cured meats with unwashed hands, followed by holding the food at warm temperatures (>40°F), can result in bacterial growth and toxin formation. While temperatures of 120°F can kill the bacterium itself, its toxin is heat resistant; therefore, it is important to keep the *Staphylococcus* organism from growing in foods. Use proper food handling practices to avoid contact with potentially contaminated surfaces and materials. Keep food either hot (above 140°F) or cold (below 40°F) during serving time, and as quickly as possible, refrigerate or freeze leftovers and foods to be served later. *Staphylococcus aureus* is destroyed by cooking and other thermal processing, but can be reintroduced via mishandling; the bacteria can then produce a toxin that is not destroyed by further cooking. Dry curing may or may not destroy *S. aureus*, but the high salt content on the exterior of dry cured meats inhibits these bacteria. When the dry cured meat is sliced, the moist, lower salt interior will permit staphylococcal multiplication.

Salmonella

Salmonella outbreaks have been recorded for raw meats, poultry, and fish and beef jerky. Salmonella bacteria thrive at temperatures between 40-140°F. They are readily destroyed by cooking to 165°F and do not grow at refrigerator or freezer temperatures. They do survive refrigeration and freezing, however, and will begin to grow again once warmed to room temperature.

Campylobacter

Raw chicken is a primary source of this organism, which grows best in a reduced oxygen environment. It is easily killed by heat (120°F), is inhibited by acid, salt and drying, and will not multiply at temperatures below 85°F. *Campylobacter* is the leading bacterial cause of diarrhea in the U.S.

Vibrio

Infections with this organism have been associated with the consumption of raw, improperly cooked, or cooked and recontaminated fish and shellfish. A correlation exists between the probability of infection and warmer months of the year. Improper refrigeration of seafood contaminated with this organism will allow its proliferation, increasing the possibility of infection. People with liver disease are particularly at risk for infection caused by undercooked seafood containing *V. vulnificus*.

Parasites (Other than Trichinella)

Anisakis simplex parasites are known to occur frequently in the flesh of cod, haddock, fluke, pacific salmon, herring, flounder, and monkfish. However, only 10 reported cases annually in the U.S. are attributed to them. *Diphyllobothrium latum* and *Nanophyetus*spp. parasites are known to occur frequently in the flesh of fish. Foodborne illnesses attributed to them are few in number. Sufficient cooking of foods would destroy the parasites.

In the Great Lakes region of the U. S., the Broad Fish Tapeworm has resulted in food poisoning outbreaks related to pickled pike. The larvae pass through small fish until they hatch as small worms in larger fish. If consumed at this stage by humans the worms can grow in the intestines. Sufficient cooking of foods would destroy the parasites.

Viruses

Shellfish are the food most often implicated foods in outbreaks of viruses such as Norwalk and Hepatitis A. Ingestion of raw or insufficiently steamed clams and oysters poses a high risk for infection with viruses. Sufficient cooking of foods would destroy the viruses.

Inhibition of Pathogens in Cured Meats

Salt and nitrates or nitrites are the primary chemicals that are responsible for the inhibition of pathogen growth when curing meats. Adding to that, pH and temperature (below 40°F or above 140°F), these factors can act in concert to prohibit the growth of pathogens in these foods. Table indicates some extreme parameters for growth of pathogens.

Organism	min. pH	max. % salt	min. temp.	oxygen req.
Campylobacter	4.9	2	86°F	MA1
Clostridium	4.7	10	38°F	AN2
E. coli	3.6	8	33°F	FA3
Listeria	4.8	12	32°F	FA
Salmonella	4.0	8	41°F	FA
Staphylococcus	4.0	20	41°F	FA
Vibrio	3.6	10	41°F	FA
1MA=microaerophilic; requires limited levels of oxygen; 2AN=anaerobic, requires the absence of oxygen; and 3FA=facultative anaerobic, can grow either with or without oxygen.				

Critical Parameters for growth of some Pathogens.

Cured/Smoked Food Poisoning

Ham

Trichinella, Staphylococcus, and molds are the microorganisms most associated with ham. All ham should be processed to specifically kill trichinae. *Staphylococcus aureus,* which is salt tolerant, can, survive the high salt levels of the ham surface. Once the ham is sliced, *S. aureus* can grow on the interior tissues where there is a lower salt concentration. Therefore, the USDA-FSIS recommends that all sliced ham be refrigerated. Molds can grow on the ham surface, especially on country-cured hams. The USDA-FSIS recommends that you wash the ham free of the mold with a stiff vegetable brush and that consumption of the ham is safe. We were unable to find any studies of aflatoxin formation with molds associated with hams.

Bacon

Like other cured products, *Listeria monocytogenes* has been responsible for a number of recalls of ready-to-eat bacon, e.g., State of Ohio Department of Agriculture Recall Announcement (ODA/ODH) 99 05a. Packages stored at room temperature sampled positive for the pathogen.

Beef

Pastrami made in a small Idaho commercial firm tested positive for *Listeria monocytogenes* in July 2000. No reports of food poisonings were recorded, but the products were recalled.Corned beef samples also tested positive for *Listeria monocytogenes* from a Michigan commercial firm. Corned beef was cooked and temperature abused at a deli in Ohio resulting in an outbreak of *C. perfringens* food poisoning.

Poultry

Much of the reports of food poisoning and recalls of poultry products for have been with commercial ready to eat products, such as chicken or turkey lunchmeats.

Fish

Listeria monocytogenes has been found in commercial samples of cold smoked fish leading to product recalls in New York (Cold smoked sea bass FDA Recall No.F-313-1) and Seattle, WA (Cold smoked salmon FDA Recall #F-265-1). These recalls demonstrate that even with HACCP and careful plant sanitation, commercial processors have contamination incidences in their cold smoked fish processes. In New York, fish sausage was recalled because laboratory analysis found pH (acidity), salt and water activity levels in the product were such that they could potentially permit *Clostridium botulinum* to develop and produce the toxin.

Sausage

Recent concern about the safety of sausages has been in the semi-dry fermented sausages, such as summer sausage. *E. coli* O157:H7 has been found to survive the acidity of these products. Some commercial, ready-to-eat sausages and luncheon meats have been implicated in *Listeria monocytogenes* growth and outbreaks. Additional concerns with trichinae may occur in any pork sausage.

Game

Precaution should be used since venison, bear, elk, wild boar, wild turkey, rabbit and other game animals are usually field dressed in unknown sanitary conditions or kept from immediate refrigeration. Two areas of special interest should be noted: (1) *E. coli*O157:H7 outbreaks in game sausage and jerky, and (2) Trichinosis in game meats from northern U.S. areas. Several outbreaks of *E. coli*O157:H7 have occurred in venison jerky.

T. nativa is an Alaskan, Canadian, and Arctic strain of *Trichinella* that is freeze-resistant. Unlike pork, freezing arctic meat will not kill larval cysts. Wild game, e.g., bear or walrus meat, is safe once the entire piece is completely cooked. USDA recommends attaining an internal temperature of at least 170°F. Since cooking may be uneven, microwaving of game meats is not recommended,

Cured/Smoked Food Spoilage

Not all microbial growth leads to food poisoning. Indeed, many organisms simply spoil cured and smoked foods making them unpalatable. Keep in mind that it is a general rule that if conditions exist to allow growth of spoilage organisms, these same conditions can allow for the growth of food poisoning organisms. Good judgment should prevail.

Lactic Acid Bacteria

Lactic acid bacteria are frequent spoilage organisms on cured/smoked meats. They are tolerant of some of the conditions in the curing/smoking process or are contaminates after processing. They grow slowly, but eventually spoil the food by producing organic acids.

Mold and Cured Meats

Moldy cured or smoked meat is a controversial topic. Very often country hams will have a moldy surface. Currently the USDA FSIS recommends cleaning the mold and soaking the ham in water to refresh it is a safe procedure. Other suggestions are to wash the ham in acetic acid (acetic acid Avinegar@ 10% in water).

Greening of Cured/Smoked Meats

Lactobacillus viridescens, or similar bacteria that produce hydrogen peroxide may cause greening in meats. The H_2O_2 reacts with myoglobin to produce a green sheen pigment. The meat, while less appealing, is not dangerous to consume.

Slime Producers

Some *Micrococcus* spp. and other bacteria are capable of producing slime on the surface of hams, bacon, and sausages.

Gas Producers

Some organisms can produce gas pockets inside cured or smoked meats.

Rancid Flavors in Home Cured Pork

Salt increases oxidation during long cures and can lead to a rancid flavor. Prolonged frozen storage may also contribute to oxidation leading to rancid flavors. Many consumers prefer these flavors. For those that do not, shorter curing and aging times should be considered.

Cooling

Cooling can be defined as a processing technique that is used to reduce the temperature of the food from one processing temperature to another or to a required storage temperature. Chilling is a processing technique in which the temperature of a food is reduced and kept at a temperature between −1°C and 8°C. The objective of cooling and chilling is to reduce the rate of biochemical and microbiological changes in foods, in order to extend the shelf-life of fresh and processed foods, or to maintain a certain temperature in a food process, e.g. in the fermentation and treatment of beer. Cooling is also used to promote a change of state of aggregation, e.g. crystallisation. In the wine industry, cooling (chilling) is applied to clarify the must before fermentation. The objective of cold stabilisation is to obtain the precipitation of tartrates (in wines) or fatty acids (in spirits) before bottling.

Applications

Cooling is a process step in many food production processes. Chilling for food preservation is widely applied for a lot of perishable foods. The main application of cold stabilisation in the food industry is in the wine and spirit sector.

The supply of chilled foods to consumers requires a sophisticated distribution system, involving chilled stores, refrigerated transport and chilled retail display cabinets. Chilled foods can be grouped into three categories according to the storage temperature (Hendley, B. (1985) Market for chilled foods. Food Process 52, 29-33) and the forth is applied in wine making:

- −1°C to + 1°C (fresh fish, meats, sausages and ground meats, smoked meats and fish);

- 0°C to + 5°C (pasteurised canned meat, milk and milk products, prepared salads, baked goods, pizzas, unbaked dough and pastry);

- 0°C to + 8°C (fully cooked meats and fish pies, cooked or uncooked cured meats, butter, margarine, cheese and soft fruits);

- 8°C to 12°C in the wine industry. The must is kept at this temperature between 6 and 24 hours.

Techniques, Methods and Equipments

Cooling of liquid foods is commonly carried out by passing the product through a heat exchanger (cooler) or by cooling the vessels. The cooling medium in the cooler can be groundwater, water recirculating over a cooling tower, or water (eventually mixed with agents like glycol) which is recirculated via a mechanical refrigeration system (ice-water). Cooling and chilling of solid foods is carried out by contacting the food with cold air, or directly with a refrigerant like liquid carbon dioxide or liquid nitrogen. The equipment used for freezing can also be used for cooling and chilling.

Some typical applications are given below:

a) Cooling of sugar

   Sugar to be stored in silos must be dedusted and cooled to the storage temperature. This is done in a sugar cooler, which is a device in which warm and dried sugar is intensively aerated by cold filtered external air to cool the sugar to the storage temperature, approximately 20–30°C. The most common systems in use are coolers (typically drum or fluidised-bed coolers) with chilling systems with countercurrent or cross-current phase flow.

b) Cryogenic cooling

   In cryogenic cooling the food is in direct contact with the refrigerant, which may be solid or liquid carbon dioxide, or liquid nitrogen. As the refrigerant evaporates or sublimates it removes heat from the food, thereby causing rapid cooling. Both liquid nitrogen and carbon dioxide refrigerants are colourless, odourless and inert.

c) Cold stabilisation

   Cold stabilisation is a technique of chilling wines before bottling to cause the precipitation of harmless tartrate crystals.

   For spirits, this technique consists of bringing the spirit to a temperature of between -1°C and -7°C, depending on the operators, and possibly performing a stabulation (storing at low temperature) in a tank at constant temperature for between 24 and 48 hours. A cold filtration (around -1°C) allows the fatty acid esters to be retained.

For wines, three techniques can be employed: stabilisation by batch and stabulation. This is the oldest technique and consists of bringing the wine to a temperature below zero close to the freezing point, then stabulating in an isothermal tank during a period of 5 to 8 days.

But currently the most widely-used techniques are: continuous stabilisation, where the stabulation tank is replaced by a cylindro-conical crystalliser and an agitator, in which the wine will remain for only between 30 and 90 minutes, stabilisation by crystal seeding consisting of refrigerating at between -1° and -2°C, and seeding at 4 g/l of tartaric crystals with agitation over 2 to 4 hours, and later storage in tank, and decantation after 12 to 48 hours. There can be many variations on these basic schemes.

Freezing

Freezing is one of the oldest and most widely used methods of food preservation, which allows preservation of taste, texture, and nutritional value in foods better than any other method. The freezing process is a combination of the beneficial effects of low temperatures at which microorganisms cannot grow, chemical reactions are reduced, and cellular metabolic reactions are delayed.

The Importance of Freezing as a Preservation Method

Freezing preservation retains the quality of agricultural products over long storage periods. As a method of long-term preservation for fruits and vegetables, freezing is generally regarded as superior to canning and dehydration, with respect to retention in sensory attributes and nutritive properties. The safety and nutrition quality of frozen products are emphasized when high quality raw materials are used, good manufacturing practices are employed in the preservation process, and the products are kept in accordance with specified temperatures.

The Need for Freezing and Frozen Storage

Freezing has been successfully employed for the long-term preservation of many foods, providing a significantly extended shelf life. The process involves lowering the product temperature generally to -18 °C or below. The physical state of food material is changed when energy is removed by cooling below freezing temperature. The extreme cold simply retards the growth of microorganisms and slows down the chemical changes that affect quality or cause food to spoil.

Competing with new technologies of minimal processing of foods, industrial freezing is the most satisfactory method for preserving quality during long storage periods. When compared in terms of energy use, cost, and product quality, freezing requires the shortest processing time. Any other conventional method of preservation focused on fruits and vegetables, including dehydration and canning, requires less energy when compared with energy consumption in the freezing process and storage. However, when the overall cost is estimated, freezing costs can be kept as low (or lower) as any other method of food preservation.

Current Status of Frozen Food Industry in U.S. and other Countries

The frozen food market is one of the largest and most dynamic sectors of the food industry. In spite of considerable competition between the frozen food industry and other sectors, extensive quantities of frozen foods are being consumed all over the world. The industry has recently grown to a value of over US$ 75 billion in the U.S. and Europe combined. This number has reached US$ 27.3 billion in 2001 for total retail sales of frozen foods in the U.S. alone (AFFI, 2003). In Europe, based on U.S. currency, frozen food consumption also reached 11.1 million tons in 13 countries in the year 2000. represents the division of frozen food industry in terms of annual sales in 2001.

Advantages of Freezing Technology in Developing Countries

Developed countries, mostly the U.S., dominate the international trade of fruits and vegetables. The U.S. is ranked number one as both importer and exporter, accounting for the highest percent of fresh produce in world trade. However, many developing countries still lead in the export of fresh exotic fruits and vegetables to developed countries.

For developing countries, the application of freezing preservation is favorable with several main considerations. From a technical point of view, the freezing process is one of the most convenient and easiest of food preservation methods, compared with other commercial preservation techniques. The availability of different types of equipment for several different food products results in a flexible process in which degradation of initial food quality is minimal with proper application procedures. As mentioned earlier, the high capital investment of the freezing industry usually plays an important role in terms of economic feasibility of the process in developing countries. As for cost distribution, the freezing process and storage in terms of energy consumption constitute approximately 10 percent of the total cost. Depending on the government regulations, especially in developing countries, energy cost for producers can be subsidized by means of lowering the unit price or reducing the tax percentage in order to enhance production. Therefore, in determining the economical convenience of the process, the cost related to energy consumption (according to energy tariffs) should be considered.

Food items	Sales US$ (million)	% Change vs. 2000
Total Frozen Food Sales	26 600	6.1
Baked Goods	1 400	9.0
Breakfast Foods	1 050	4.1
Novelties	1 900	10.5
Ice Cream	4 500	5.7
Frozen Dessert/Fruit/Toppings	786	5.4
Juices/Drinks	827	-9.7
Vegetables	2 900	4.3

Frozen food industry in terms of annual sales in 2001

Increasing Consumer Demand in Developing Countries due to Modernization

The proportion of fresh food preserved by freezing is highly related to the degree of economic development in a society. As countries become wealthier, their demand for high-valued commodities increases, primarily due to the effect of income on the consumption of high-valued commodities in developing countries. The commodities preserved by freezing are usually the most perishable ones, which also have the highest price. Therefore, the demand for these commodities is less in developing areas. Besides, the need for adequate technology for freezing process is the major drawback of developing countries in competing with industrialized countries. The frozen food industry requires accompanying developments and facilities for transporting, storing, and marketing their products from the processing plant to the consumer. Thus, a large amount of capital investment is needed for these types of facilities. For developing countries, especially in rural or semi-rural areas, the frozen food industry has therefore not been developed significantly compared to other countries.

In recent years, due to the changing consumer profile, the frozen food industry has changed significantly. The major trend in consumer behavior documented over the last half century has been the increase in the number of working women and the decline in the family size. These two factors resulted in a reduction in time spent preparing food. The entry of more women into the workforce also led to improvements in kitchen appliances and increased the variability of ready-to-eat or frozen foods available in the market. Besides, the increased usage of microwave ovens, affecting food habits in general and the frozen food market in particular, as well as allowing rapid preparation of meals and greater flexibility in meal preparation. The frozen food industry is now only limited by imagination, an output of which increases continuously to supply the increasing demand for frozen products and variability.

Country	1999	2000	2001	2002
Argentina	n.a.	0.075	0.069	n.a.
Belgium	0.056	0.048	n.a.	n.a.
Bolivia	n.a.	0.062	0.069	n.a.
Chile	n.a.	0.052	0.056	n.a.
Chinese Taipei (Taiwan)	0.058	0.061	0.056	n.a.
Colombia	n.a.	0.052	0.042	n.a.
Costa Rica	n.a.	0.068	0.076	n.a.
Cuba	n.a.	0.080	0.078	n.a.
Ecuador	n.a.	0.036	0.061	n.a.
El Salvador	n.a.	0.111	0.110	n.a.
Finland	0.046	0.039	0.038	0.043
Germany	0.057	0.041	0.044	n.a.
Greece	0.050	0.042	0.043	0.046
Guyana	n.a.	0.082	0.080	n.a.
Hungary	0.055	0.049	0.051	0.060
India	0.081	0.080	n.a.	n.a.
Ireland	0.057	0.049	0.060	0.075
Italy	0.086	0.089	n.a.	n.a.

Korea (Korea, South)	0.056	0.062	0.057	n.a.
Mexico	0.042	0.051	0.053	n.a.
Netherlands	0.061	0.057	0.059	n.a.
New Zealand	0.030	0.030	0.028	0.033
Nicaragua	n.a.	0.117	0.115	n.a.
Paraguay	n.a.	0.032	0.036	n.a.
Peru	n.a.	0.056	0.057	n.a.
Poland	0.037	0.037	0.045	0.049
Portugal	0.078	0.067	0.066	0.068
Russia	0.012	0.011	n.a.	n.a.
South Africa	0.017	0.017	0.013	n.a.
Spain	0.049	0.043	0.041	n.a.
Switzerland	0.090	0.069	0.069	0.073
Turkey	0.079	0.080	0.079	0.094
United Kingdom	0.064	0.055	0.048	n.a.
United States 2	0.044	0.046	0.050	0.048
Uruguay	n.a.	0.064	0.070	n.a.

Unit electricity prices for industry1 (U.S. Dollars per Kilowatt-hour)

n.a. = Not Available.

- Energy end-use prices including taxes converted using exchange rates

- Electricity prices in the United States, including income taxes, environmental charges, and other charges.

Market Share of Frozen Fruits and Vegetables

Today in modern society, frozen fruits and vegetables constitute a large and important food group among other frozen food products. The historical development of commercial freezing systems designed for special food commodities helped shape the frozen food market. Technological innovations as early as 1869 led to the commercial development and marketing of some frozen foods. Early products saw limited distribution through retail establishments due to insufficient supply of mechanical refrigeration. Retail distribution of frozen foods gained importance with the development of commercially frozen vegetables in 1929.

The frozen vegetable industry mostly grew after the development of scientific methods for blanching and processing in the 1940s. Only after the achievement of success in stopping enzymatic degradation, did frozen vegetables gain a strong retail and institutional appeal. Today, market studies indicate that considering overall consumption of frozen foods, frozen vegetables constitute a very significant proportion of world frozen-food categories (excluding ice cream) in Austria, Denmark, Finland, France, Germany, Italy, Netherlands, Norway, Sweden, Switzerland, UK, and the USA. The division of frozen vegetables in terms of annual sales in 2001.

Commercialization history of frozen fruits is older than frozen vegetables. The commercial freezing of small fruits and berries began in the eastern part of the U.S. in about 1905. The main advantage of freezing preservation of fruits is the extended usage of frozen fruits during off-season.

Additionally, frozen fruits can be transported to remote markets that could not be accessed with fresh fruit. Also, freezing preservation makes year-round further processing of fruit products possible, such as jams, juice, and syrups from frozen whole fruit, slices, or pulps. In summary, the preservation of fruits by freezing has clearly become one the most important preservation methods.

Future Trends in Freezing Technology

The frozen food industry is highly based in modern science and technology. Starting with the first historical development in freezing preservation of foods, today, a combination of several factors influences the commercialization and usage of freezing technology. The future growth of frozen foods will mostly be affected by economical and technological factors. Growth in population, personal incomes, relative cost of other forms of foods, changes in tastes and preferences, and technological advances in freezing methods are some of the factors concerned with the future of freezing technology.

Population growth and increasing demand for food has generated the need for commercial production of food commodities in large-scale operations. Thus, availability of proper equipment suitable for continuous processing would be valuable for freezing preservation methods. In addition depending on personal incomes, relative cost of frozen products is one of the most important of economical factors. Producing the highest quality at the lowest cost possible is highly dependent on the technology used. As a result, developments in freezing technology in recent years have mostly been characterized by the improvements in mechanical handling and process control to increase freezing rate and reduce cost.

Today an increasing demand for frozen foods already exits and further expansion of the industry is primarily dependent on the ability of food processors to develop higher qualities in both process techniques and products. Improvements can only be achieved by focusing on new technologies and investigating poorly understood factors that influence the quality of frozen food products. Improvements in new and convenient forms of foods, as well as more information on relative cost and nutritive values of frozen foods, will contribute toward continued growth of the industry.

Vegetables	Sales US$ (million)	% Change vs. 2000
Broccoli	184	4.4
Com/Corn on the Cob	312	3.5
Green Beans	115	6.0
Mixed Vegetables	450	7.2
Peas	207	3.9
Potatoes	1 070	4.4

Frozen vegetables in terms of annual sales in 2001

General Recommendations on the Freezing Process

Freezing is a widely used method of food preservation based on several advantages in terms of retention of food quality and ease of process. Beginning with the earliest history of freezing, the technology has been highly affected over the years by the developments and improvements in freezing techniques. In order to understand and handle the concepts associated with freezing of foods, it is necessary to examine the fundamental factors governing the freezing process.

Freezing Technology

Freezing has long been used as a method of preservation, and history reveals it was mostly shaped by the technological developments in the process. A small quantity of ice produced without using a "natural cold" in 1755 was regarded as the first milestone in the freezing process. Firstly, ice-salt systems were used to preserve fish and later on, by the late 1800's, freezing was introduced into large-scale operations as a method of commercial preservation. Meat, fish, and butter, the main products preserved in this early example, were frozen in storage chambers and handled as bulk commodities.

In the following years, scientists and researchers continuously worked to achieve success with commercial freezing trials on several food commodities. Among these commodities, fruits were one of the most important since freezing during the peak growing season had the advantage of preserving fruit for later processing into jams, jellies, ice cream, pies, and other bakery foods. Although commercial freezing of small fruits and berries first began around 1905 in the eastern part of the United States, the commercial freezing of vegetables is much more recent. Starting from 1917, only private firms conducted trials on freezing vegetables, but achieving good quality in frozen vegetables was not possible without pre-treatments due to the enzymatic deterioration. In 1929, the necessity of blanching to inactivate enzymes before freezing was concluded by several researchers to avoid deterioration and off-flavours caused by enzymatic degradation.

The modern freezing industry began in 1928 with the development of double-belt contact freezers by a technologist named Clarence Birdseye. After the revolution in the quick freezing process and equipment, the industry became more flexible, especially with the usage of multi-plate freezers. The earlier methods achieved successful freezing of fish and poultry, however with the new quick freezing system, packaged foods could be frozen between two metal belts as they moved through a freezing tunnel. This improvement was a great advantage in the commercial large-scale freezing of fruits and vegetables. Furthermore, quick-freezing of consumer-size packages helped frozen vegetables to be accepted rapidly in late 1930s.

Today, freezing is the only large-scale method that bridges the seasons, as well as variations in supply and demand of raw materials such as meat, fish, butter, fruits, and vegetables. Besides, it makes possible movement of large quantities of food over geographical distances. It is important to control the freezing process, including the pre-freezing preparation and post-freezing storage of the product, in order to achieve high-quality products. Therefore, the theory of the freezing process and the parameters involved should be understood clearly.

Freezing Process

The freezing process mainly consists of thermodynamic and kinetic factors, which can dominate each other at a particular stage in the freezing process. Major thermal events are accompanied by reduction in heat content of the material during the freezing process as is shown in figure. The material to be frozen first cools down to the temperature at which nucleation starts. Before ice can form, a nucleus, or a seed, is required upon which the crystal can grow; the process of producing this seed is defined as nucleation. Once the first crystal appears in the solution, a phase change occurs from liquid to solid with further crystal growth. Therefore, nucleation serves as the initial process of freezing, and can be considered as the critical step that results in a complete phase change.

Freezing Point of Foods

Freezing point is defined as the temperature at which the first ice crystal appears and the liquid at that temperature is in equilibrium with the solid. If the freezing point of pure water is considered, this temperature will correspond to 0°C (273°K). However, when food systems are frozen, the process becomes more complex due to the existence of both free and bound water. Bound water does not freeze even at very low temperatures. Unfreezable water contains soluble solids, which cause a decrease in the freezing point of water lower than 0°C. During the freezing process, the concentration of soluble solids increases in the unfrozen water, resulting in a variation in freezing temperature. Therefore, the temperature at which the first ice crystal appears is commonly regarded as the initial freezing temperature. There are empirical equations in literature that can calculate the initial freezing temperature of certain foods as a function of their moisture content.

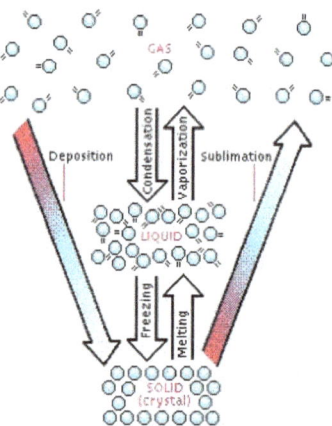

A schematic illustration of overall freezing process.

There are several methods of food freezing, and depending on the method used, the quality of the frozen food may vary. However, regardless of the method chosen, the main principle behind all freezing processes is the same in terms of process parameters. The International Institute of Refrigeration (IIR) has provided definitions to establish a basis for the freezing process. According to their definition, the freezing process is basically divided into three stages based on major temperature changes in a particular location in the product, as shown in figures for pure water and food respectively.

Beginning with the prefreezing stage, the food is subjected to the freezing process until the appearance of the first crystal. If the material frozen is pure water, the freezing temperature will be 0 °C and, up to this temperature, there will be a sub cooling until the ice formation begins. In the case of foods during this stage, the temperature decreases to below freezing temperature and, with the formation of the first ice crystal, increases to freezing temperature. The second stage is the freezing period; a phase change occurs, transforming water into ice. For pure water, temperature at this stage is constant; however, it decreases slightly in foods, due to the increasing concentration of solutes in the unfrozen water portion. The last stage starts when the product temperature reaches the point where most freezable water has been converted to ice, and ends when the temperature is reduced to storage temperature.

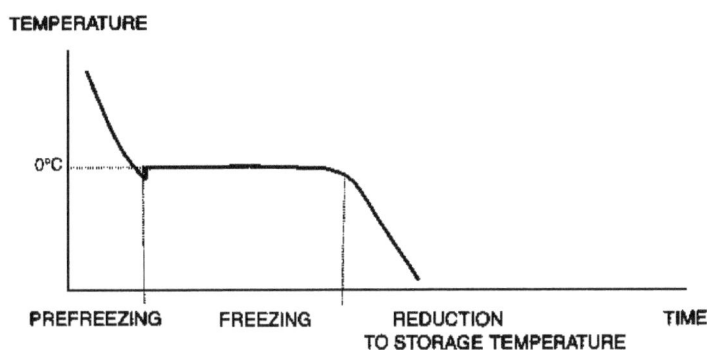

Practical definition of the freezing process for pure water.

The freezing time and freezing rate are the most important parameters in designing freezing systems. The quality of the frozen product is mostly affected by the rate of freezing, while time of freezing is calculated according to the rate of freezing. For industrial applications, they are the most essential parameters in the process when comparing different types of freezing systems and equipment.

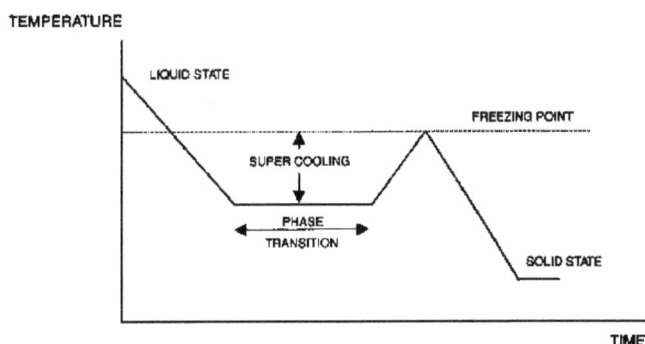

Practical definition of the freezing process for foods.

Freezing Time

Again, freezing time is one of the most important parameters in the freezing process, defined as time required to lower product temperature from its initial temperature to a given temperature at its thermal center. Since the temperature distribution within the product varies during freezing process, the thermal center is generally taken as reference. Thus, when the geometrical center of the product reaches the given final temperature, this ensures the average product temperature has been reduced to a storage value. Freezing time depends on several factors, including the initial and final temperatures of the product and the quantity of heat removed, as well as dimensions (especially thickness) and shape of product, heat transfer process, and temperature. The International Institute of Refrigeration (1986) defines various factors of freezing time in relation to both the product frozen and freezing equipment. The most important are:

- Dimensions and shape of product, particularly thickness;

- Initial and final temperatures;

- Temperature of refrigerating medium;

- Surface heat transfer coefficient of product;

- Change in enthalpy;

- Thermal conductivity of product.

Calculation of freezing time in food systems is difficult in comparison to pure systems since the freezing temperature changes continuously during the process. Using a simplified approach, time elapsed between initial freezing until when the entire product is frozen can be regarded as the freezing time. Plank's equation is commonly used to estimate freezing time, however due to assumptions involved in the calculation it is only useful for obtaining an approximation of freezing time. The derivation of the equation starts with the assumption the product being frozen is initially at freezing temperature. Therefore, the calculated freezing time represents only the freezing period. The equation can be further modified for different geometries including slab, cylinder, and sphere, where for each geometry, the coefficients are arranged in relation to the dimensions.

Geometry	P	R	Dimension
Infinite slab	1/2	1/8	thickness e
Infinite cylinder	1/4	1/16	radius r
Sphere	1/6	1/24	radius r

Coefficients P and R of Equation 1

$$t_F = \frac{\rho H}{T_F - T_e} \left[\frac{e^2 R}{k} + \frac{eP}{h} \right]$$

where l_1 is the latent heat of frozen fraction, k and r are the thermal conductivity and density of the frozen layer, while h is the coefficient of heat transfer by convection to the exterior. T_f denotes the body temperature of the product when introduced into a freezer in wich the external temperature is T_e The coefficients R and P are given in table and arranged according to the geometry of the product frozen. where the letter e denotes the dimension (i.e. for infinite slab geometry, e is thickness of the slab and for infinite cylinder or sphere e is replaced by r which denotes the radius of the clylinder or sphere).

As mentioned earlier, the equation of Plank assumes the food is at a freezing temperature at the beginning of the freezing process. However, the food is usually at a temperature higher than freezing temperature. The real freezing time should therefore be the sum of time calculated from the equation of Plank and the time needed for the product's surface to decrease from initial temperature to freezing temperature.

Several works have attempted to calculate real freezing time, as in one presented by Nagaoka *et al*. Nagaoka's equation calculates the amount of heat elimination required to decrease a product's temperature from initial temperature to freezing temperature, as well as the amount of heat released during the phase change and the amount of heat eliminated to reach freezing temperature. Further empirical equations can be found in literature in detail.

$$t_F = \frac{\rho \Delta H}{T_F - T_e}\left[\frac{Re^2}{k} + \frac{PI}{h}\right]\left[1 + 0.008\left(T_i - T_F\right)\right]$$

Where, T_i is the temperature of the food at the initiation of freezing, DH is the difference between the enthalpy of the food at initial temperature and end of freezing. Re and Pl are the dimensionless numbers, while k and h are the thermal conductivity and the coefficient of heat transfer, respectively.

For calculating freezing time of products with irregular shape, a common property of most food products - especially fruits and vegetables - a dimensionless factor has been employed in equations.

Freezing Rate

The freezing rate (°C/h) for a product or package is defined as the ratio of difference between initial and final temperature of product to freezing time. At a particular location within the product, a local freezing rate can be defined as the ratio of the difference between the initial temperature and desired temperature to the time elapsed in reaching the given final temperature. The quality of frozen products is largely dependent on the rate of freezing (Ramaswamy and Tung, 1984). Generally, rapid freezing results in better quality frozen products when compared with slow freezing. If freezing is instantaneous, there will be more locations within the food where crystallization begins. In contrast, if freezing is slow, the crystal growth will be slower with few nucleation sites resulting in larger ice crystals. Large ice crystals are known to cause mechanical damage to cell walls in addition to cell dehydration. Thus, the rate of freezing for plant tissues is extremely important due to the effect of freezing rate on the size of ice crystals, cell hydration, and damage to cell walls. The figure shows a general behavior of the dynamics curve of freezing preservation.

Rapid freezing is advantageous for freezing of many foods, however some products are susceptible to cracking when exposed to extremely low temperature for long periods. Several mechanisms, including volume expansion, contraction and expansion, and building of internal pressure, are proposed in literature explaining the mechanisms of product damage during freezing.

Energy Requirements

For fruits and vegetables, the amount of energy required for freezing is calculated based on the enthalpy change and the amount of product to be frozen. The following equation is reported by Riedel (1949) for calculation of refrigeration requirements for fruits and vegetables.

$$\Delta H = \left[1 - \frac{X_{SNJ}}{100}\right]\Delta H_j + 1.21\left[\frac{X_{SNJ}}{100}\right]\Delta T$$

X_{SNJ} : Percentage of the product solids different from juice (Dry matter fraction of the juice);

DH$_j$: Enthalpy change during freezing of the juice fraction;

DT : Temperature difference between initial and final temperature of the product.

Refrigeration

Refrigeration is defined as the elimination of heat from a material at a temperature higher than the temperature of its surroundings. The mechanism of refrigeration is a part of the freezing process and freezing storage involved in the thermodynamic aspects of freezing. According to the second law of thermodynamics, heat only flows from higher to lower temperatures. Therefore, in order to raise the heat from a lower to a higher temperature level, expenditure of work is needed. The aim of industrial refrigeration processes is to eliminate heat from low temperature points towards points with higher temperature. For this reason, either closed mechanical refrigeration cycles in which refrigeration fluids circulate, or open cryogenic systems with liquid nitrogen (LIN) or carbon dioxide (CO_2), are commonly used by the food industry.

The main elements in a closed mechanical refrigeration system are the condenser, compressor, evaporator, and the expansion valve. The refrigerants hydro chlorofluorocarbon (HCFC) and ammonia are examples of the refrigerants circulated in these types of mechanical refrigeration systems. A simple scheme for the closed mechanical refrigeration system.

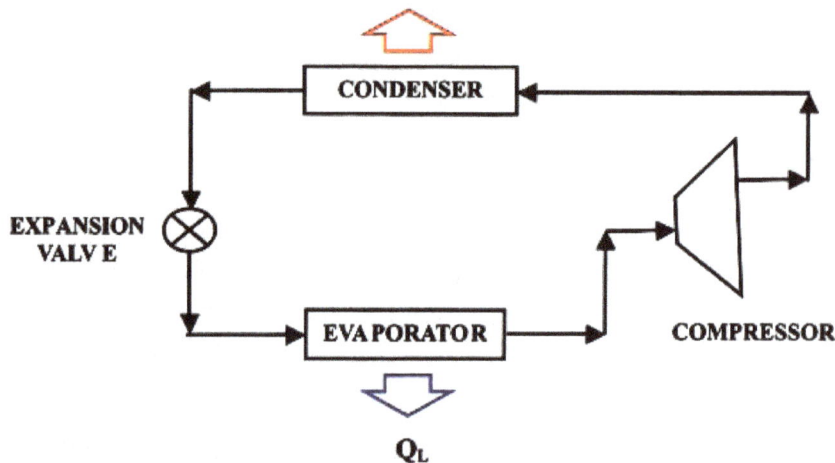

A simple scheme for a one-stage closed mechanical refrigeration system.

Starting at the suction point of the compressor, fluid in a vapor state is compressed into the compressor where an increase in temperature and pressure takes place. The fluid then flows through the condenser where it decreases in energy by giving off heat and converting to a liquid state. After the phase, a change occurs inside the condenser, the fluid flows through the expansion valve where the pressure decreases to convert liquid into a form of liquid-gas mixture. Finally, the liquid-gas mixture flows through the evaporator where it is converted into a saturated vapor state and removes heat from the environment in the process of cooling. With this last stage the loop restarts again.

The other refrigeration system employed by the food industry is the cryogenic system with carbon dioxide or liquid nitrogen. The refrigerant in this system is consumed differently from the circulating fluid in closed mechanical systems.

Refrigerants

There are several refrigerants available for refrigeration systems. The selection of a proper refrigerant is based on physical, thermodynamic, and chemical properties of the fluid. Environmental considerations are also important in refrigerant selection, since leaks within the system produce deleterious effects on the atmospheric ozone layer. Some refrigerants, including halocarbons, have been banned to avoid potential hazardous effects. For industrial applications, ammonia is commonly used, while chlorofluoromethane and tetrafluoroethane are also recommended as refrigerants.

Freezing Capacity

Freezing equipment selection is based on the requirements for freezing a certain quantity of food per hour. For any type of freezer, freezing capacity (expressed in tonnes per hour) is defined as the ratio of the quantity of the product that can be loaded into the freezer to the holding time of the product in that particular freezer. The first parameter, the amount of food product loaded into the freezer, is affected by both the dimensions of the product and the mechanical constraints of the freezer. The denominator (holding time) has an important role in freezing systems and is based on the calculation of the amount of heat removed from the product per hour, which varies depending on the type of product frozen.

Freezing Systems

There is a variety of freezing systems available for freezing, and for most products, more than one type of freezer can be used. Therefore, in selecting a freezing system initially, a cost-benefit analysis should be conducted based on three important factors: economics, functionality, and feasibility. Financial considerations mainly involve capital investment and the production cost of selected equipment. Product losses during freezing operation should be included in cost estimation since generating higher cost freezers may have other benefits in terms of reducing product losses. Functional factors are mostly based on the suitability of the selected freezer for particular products. The mode of process, either in-line or batch, should be considered based on the fact that computerized systems are becoming more important for ease of handling and lowering production costs. Mechanical constraints for the freezer should also be considered since some types of freezers are not physically suitable for freezing certain products. Lastly, the feasibility of the process should be considered in terms of plant location or location of the processing area, as well as cleanability and hygienic design, and desired product quality.

These factors and initial considerations can help eliminate several choices in freezer selection, but the relative importance of factors may change depending on the process. For developing countries where the freezing application is relatively new, the cost factor becomes more important than other factors due to the decreased production rates and need for lower capital investment costs.

Freezing Equipment

The industrial equipment for freezing can be categorized in many ways, namely as equipment used for batch or in-line operation, heat transfer systems (air, contact, cryogenic), and product stability. The rate of heat transfer from the freezing medium to the product is important in defining the freezing time of the product. Therefore, the equipment selected for freezing process characterizes the rate of freezing.

Air-blast Freezers

The air blast freezer is one the oldest and commonly used freezing equipment due to its temperature stability and versatility for several product types. In general, air is used as the freezing medium in the freezing design, either as still air or forced air. Freezing is accomplished by placing the food in freezing rooms called sharp freezers. Still, air freezing is the cheapest way of freezing and has the added advantage of a constant temperature during frozen storage, which allows usage for unprocessed bulk products like beef quarters and fish. However, it is the slowest method of freezing due to the low surface heat transfer coefficient of circulating air inside the room. Freezing time in sharp freezers is largely dependent on the temperature of the freezing chamber and the type, initial temperature, and size of product. An improved version of the still air freezer is the forced air freezer, which consists of air circulation by convection inside the freezing room. However, even modification of the sharp freezer with extra refrigeration capacity and fans for increased air circulation does not help control the air flow over the products during slow freezing. A typical design for air blast freezers.

There are a considerable number of designs and arrangements for air blast freezers, primarily grouped in two categories depending on the mode of process, as either inline or batch. Continuous freezers are the most suitable systems for mass production of packaged products with similar freezing times, in which the product is carried through on trucks or on conveyors. The system works on a semi-batch principle when trucks are used, since they remain stationary during the process except when a new truck enters one end of the tunnel, thus moving the others along to release a finished one at the exit. The batch freezers are more flexible since a variety of products can be frozen at the same time on individual trolleys. Over-loading may be a problem for these types of freezers, thus the process requires closer supervision than continuous systems.

Air blast freezer.

Tunnel freezers

In tunnel freezers, the products on trays are placed in racks or trolleys and frozen with cold air circulation inside the tunnel. In order to allow air circulation, optimum space is provided between layers of trolley, which can be moved continuously in and out of the freezer manually or by forklift trucks. This freezing system is suitable for all types of products, although there are some mechanical constraints including the requirement of high manpower for handling, cleaning, and transportation of trays. A trolley for a tunnel freezer.

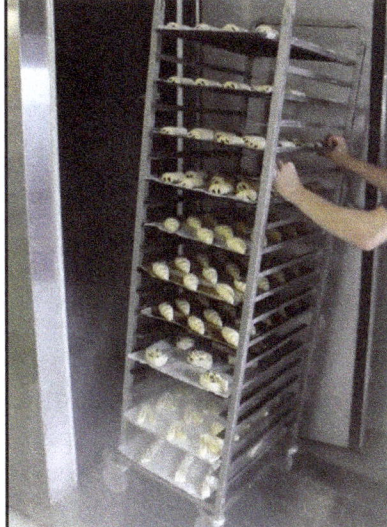

Trolley in a tunnel freezer.

Belt freezers

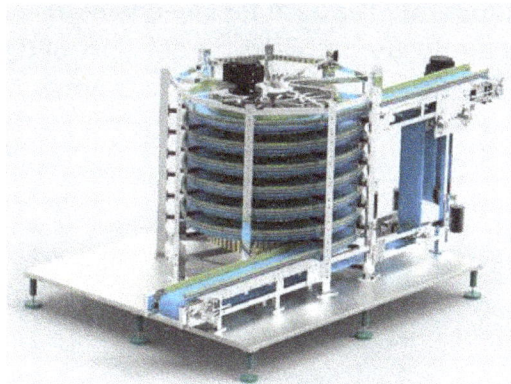

The cross-section view of a spiral belt freezer.

Belt freezers were first designed to provide continuous product flow with the help of a wire mesh conveyor inside the blast rooms. A poor heat transfer mechanism and the mechanical problems were solved in modern belt freezers by providing a vertical airflow to force air through the product layer. Airflow has good contact with the product only when the entire product is evenly distributed over the conveyor belt. In order to decrease required floor space, the belts can be arranged in a multi-tier belt freezer or a spiral belt freezer. Spiral belt freezers consist of a belt that can be bent laterally around a rotating drum to maximize belt surface area in a given floor space. This type of design has the advantage of eliminating product damage in transfer points, especially for products that require gentle handling. Both packed and unpacked products with long freezing times (10 min to 3 hr) can be frozen in spiral belt freezers due to the flexibility of the equipment. A typical spiral belt freezer.

Fluidized Bed Freezers

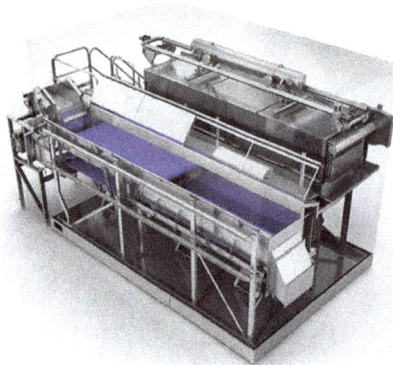

Cross-sectional view of a fluidized bed freezer.

The fluidized bed freezer, a fairly recent modified type of air-blast freezer for particular product types, consists of a bed with a perforated bottom through which cold air is blown vertically upwards. The system relies on forced cold air from beneath the conveyor belt, causing the products to suspend or float in the cold air stream. The use of high air velocity is very effective for freezing unpacked foods, especially when they can be completely surrounded by flowing air, as in the case of fluidized bed freezers.

The use of fluidization has several advantages compared with other methods of freezing since the product is individually quick frozen (IQF), which is convenient for particles with a tendency to stick together. The idea of individually quick frozen foods (IQF) started with the first technological developments aimed at quick freezing. The need for an effective means of freezing small particles with the potential for lumping during the process is the objective of IQF freezing. Small vegetables, prawns, shrimp, french-fried potatoes, diced meat, and fruits are some of the products now frozen with this technology.

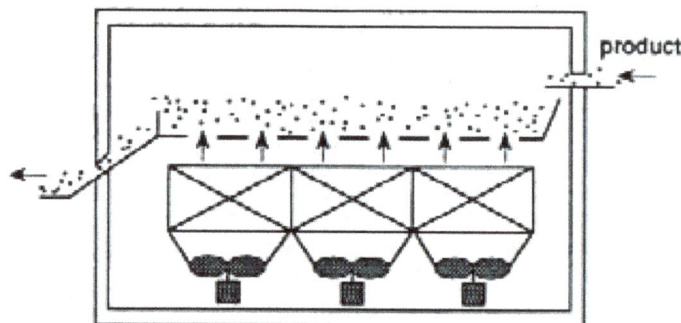

Simple working principle of a fluidized bed freezer.

Contact Freezers

Contact freezing is the one of the most efficient ways of freezing in terms of heat transfer mechanism. In the process of freezing, the product can be in direct or indirect contact with the freezing medium. For direct contact freezers, the product being frozen is fully surrounded by the freezing medium, the refrigerant, maximizing the heat transfer efficiency. A schematic illustration is given in figure. For indirect contact freezers, the product is indirectly exposed to the freezing medium while in contact with the belt or plate, which is in contact with the freezing medium.

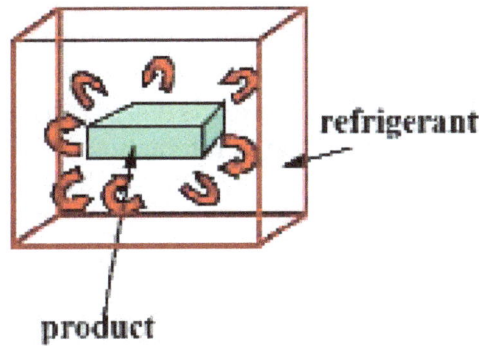

Direct contact freezer.

Immersion Freezers

The immersion freezer consists of a tank with a cooled freezing media, such as glycol, glycerol, sodium chloride, calcium chloride, and mixtures of salt and sugar. The product is immersed in this solution or sprayed while being conveyed through the freezer, resulting in fast temperature reduction through direct heat exchange. Direct immersion of a product into a liquid refrigerant is the most rapid way of freezing since liquids have better heat conducting properties than air. The solute used in the freezing system should be safe without taste, odour, colour, or flavour, and for successful freezing, products should be greater in density than the solution. Immersion freezing systems have been commonly used for shell freezing of large particles due to the reducing ability of product dehydration when the outer layer is frozen quickly. A commonly seen problem in these freezing systems is the dilution of solution with the product, which can change the concentration and process parameters. Thus, in order to avoid product contact with the liquid refrigerant, flexible membranes can be used.

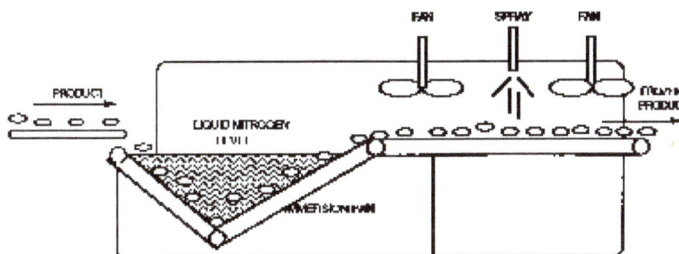

Simple illustration of a typical immersion freezer

Indirect Contact Freezers

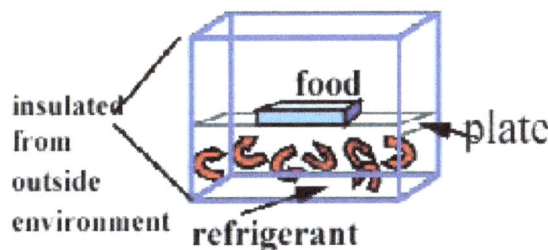

Indirect contact freezer.

In this type of freezer, materials being frozen are separated from the refrigerant by a conducting material, usually a steel plate. The mechanism of indirect contact freezer. Indirect contact freezers generally provide an efficient medium for heat transfer, although the system has some limitations, especially when used for packaged foods due to resistance of package to heat transfer. Additionally, corrosive effects may occur due to interaction of metal packages with heat transfer surfaces.

Plate Freezers

The most common type of contact freezer is the plate freezer. In this case, the product is pressed between hallow metal plates, either horizontally or vertically, with a refrigerant circulating inside the plates. Pressure is applied for good contact as schematically shown in figure.

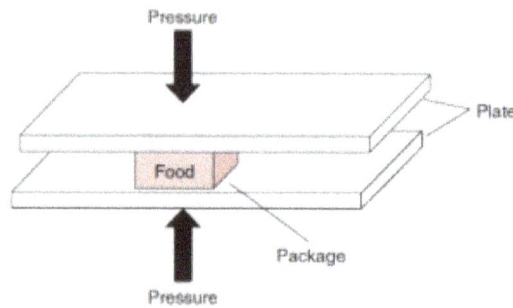

Pressure application in a plate freezer.

Plate freezer with a two-stage compressor and sea water condenser

This type of freezing system is only limited to regular-shaped materials or blocks like beef patties or block-shaped packaged products.

Contact Belt Freezers

This type of freezer is designed with single-band or double-band for freezing of thin product layers as shown in figure. The design can be either straight forward or drum. Typical products frozen in belt freezers are, fruit pulps, egg yolk, sauces and soups.

Contact belt freezer

Cryogenic Freezers

Cryogenic freezing is a relatively new method of freezing in which the food is exposed to an atmosphere below -60°C through direct contact with liquefied gases such as nitrogen or carbon dioxide. This type of system differs from other freezing systems since it is not connected to a refrigeration plant; the refrigerants used are liquefied in large industrial installations and shipped to the food-freezing factory in pressure vessels. Thus, the small size and mobility of cryogenic freezers allow for flexibility in design and efficiency of the freezing application. Low initial investment and rather high operating costs are typical for cryogenic freezers.

Liquid Nitrogen Freezers

Liquid nitrogen, with a boiling temperature of -196°C at atmospheric pressure, is a by-product of oxygen manufacture. The refrigerant is sprayed into the freezer and evaporates both on leaving the spray nozzles and on contact with the products. The system is designed in a way that the refrigerant passes in counter current to the movement of the products on the belt giving high transfer efficiency. The refrigerant consumption is in the range of 1.2-kg refrigerant per kg of the product. Typical food products used in this system are, fish fillets, seafood, fruits, berries.

Liquid Carbon Dioxide Freezers

Liquid carbon dioxide exists as either a solid or gas when stored at atmospheric pressure. When the gas is released to the atmosphere at -70°C, half of the gas becomes dry-ice snow and the other half stays in the form of vapor. This unusual property of liquid carbon dioxide is used in a variety of freezing systems, one of which is a pre-freezing treatment before the product is exposed to nitrogen spray.

Packaging

Proper packaging of frozen food is important to protect the product from contamination and damage while in transit from the manufacturer to the consumer, as well as to preserve food value, flavour, colour, and texture. There are several factors considered in designing a suitable package for a frozen food. The package should be attractive to the consumer, protected from external contamination, and effective in terms of processing, handling, and cost. Proper selection is based on the type of package and material. There are typically three types of packaging used for frozen foods: primary, secondary, and tertiary. The primary package is in direct contact with the food and the food is kept inside the package up to the time of use. Secondary packaging is a form of multiple packaging used to handle packages together for sale. Tertiary packaging is used for bulk transportation of products.

Packaging materials should be moisture-vapor-proof to prevent evaporation, thus retaining the highest quality in frozen foods. Oxygen should also be completely evacuated from the package using a vacuum or gas-flush system to prevent migration of moisture and oxygen. Glass and rigid plastic are examples of moisture-vapor-proof packaging materials. Many packaging materials, however, are not moisture-vapor-proof, but are sufficiently moisture-vapor-resistant to retain satisfactory quality in foods. Most bags, wrapping materials, and waxed cartons used in freezing packaging are moisture-vapor-resistant. In general, the containers should be leakage free while easy to seal. Durability of the material is another important factor to consider, since the packaging material must not become brittle at low temperatures and crack.

A range of different packaging materials, mainly grouped as rigid and non-rigid containers, can be used for primary packaging. Glass, plastic, tin, and heavily waxed cardboard materials are in the rigid container group and usually used for packaging of liquid food products. Glass containers are mostly used for fruits and vegetables if they are not water-packed. Plastics are the derivatives of the oil-cracking industry. Non-rigid containers include bags and sheets made of moisture-vapor-resistant heavy aluminum foil, polyethylene or laminated papers. Bags are the most commonly used packaging materials for frozen fruits and vegetables due to their flexibility during processing and handling. They can be used with or without outer cardboard cartons to protect against tearing.

Shape and size of the container are also important factors in freezing products. Serving size may vary depending on the type of product and selection should be based on the amount of food determined for one meal. For shape of the container, freezer space must be considered since rigid containers with flat tops and bottoms stack well in the freezer, while round containers waste freezer space.

Frozen Storage and Distribution

The quality of the final product depends on the history of the raw material. Using the lowest possible temperature is essential for frozen storage, transport, and distribution in achieving a high-quality product, since deteriorative processes are mainly temperature dependent. The lower the product temperature is, the slower the speed of reaction is leading to loss of quality. The temperatures of supply chains in freezing applications from the factory to the retail cabinet should be carefully monitored. The temperature regime covering the freezing process, the cold-store temperatures (£ -18°C), distribution temperatures (£ -15°C), and retail display (£ -12°C) are given as legal standards.

Freezing Fruits and Vegetables in Small and Medium Scale Operations and its Potential Applications in Warm Climates

The preservation of fruits and vegetables by freezing is one of the most important methods for retaining high quality in agricultural products over long-term storage. In particular, the freshness qualities of raw fruits and vegetables can be retained for long periods, extending well beyond the normal season of most horticultural crops. The potential application of freezing preservation of fruits and vegetables, including tropical products, has been increasing recently in parallel with developments in developing countries. Freezing of fruits and vegetables in small and medium scale operations is detailed in the following topics and a general flowchart.

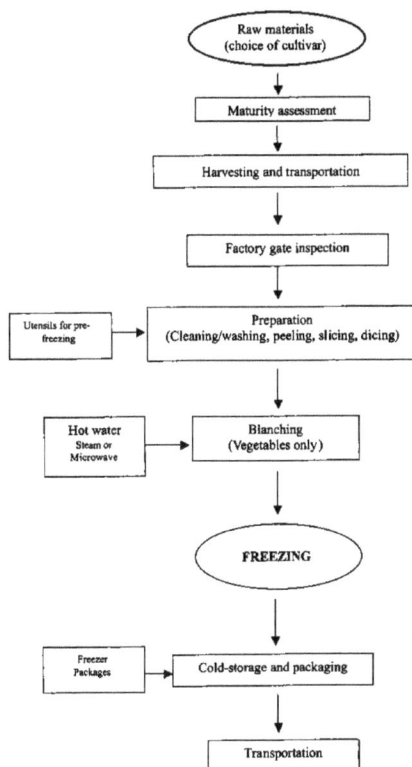

A general flow chart of frozen fruits and vegetables

Freezing Fruits

The effect of freezing, frozen storage, and thawing on fruit quality has been investigated over several decades. Today frozen fruits constitute a large and important food group. The quality demanded in frozen fruit products is mostly based on the intended use of the product. If the fruit is to be eaten without any further processing after thawing, texture characteristics are more important when compared to use as a raw material in other industries. In general, conventional methods of freezing tend to destroy the turgidity of living cells in fruit tissue. Different from vegetables, fruits do not have a fibrous structure that can resist this destructive effect. Additionally, fruits to be frozen are harvested in a fully ripe state and are soft in texture. On the contrary, a great number of vegetables are frozen in an immature state. Fruits have delicate flavours that are easily damaged or changed by heat, indicating they are best eaten when raw and decrease in quality with processing. In the same way, attractive colour is important for frozen fruits. Chemical treatments or additives are often used to inactivate the deteriorative enzymes in fruits. Therefore, proper processing is essential for all steps involved, from harvesting to packaging and distribution.

Production and Harvesting

The characteristics of raw materials are of primary importance in determining the quality of the frozen product. These characteristics include several factors such as genetic makeup, climate of the growing area, type of fertilization, and maturity of harvest.

The ability to withstand rough handling, resistance to virus diseases, molds, uniformity in ripening, and yield are some of the important characteristics of fruits in terms of economical aspects considered in production. The use of mechanical harvesting generally causes bruising of fruits and results in a wide range of maturity levels for fruits. In contrast, hand-picking provides gentler handling and maturity sorting of fruits. However in most cases, it is non-economical compared to mechanical harvesting due to high labor cost.

As a rule, harvesting of fruits at an optimum level for commercial use is difficult. Simple tests like pressure tests are applied to determine when a fruit has reached optimum maturity for harvest. Colour is also one of the characteristics used in determining maturity since increased maturation causes a darker colour in fruits. A combination of colour and pressure tests is a better way to assess maturity level for harvesting.

Controlled atmosphere storage is a common method of storage for some fruits prior to freezing. In principle, a controlled atmosphere high in carbon dioxide and low in oxygen content slows down the rate of respiration, which may extend shelf life of any respiring fruit during storage. Due to the fact that these fruits do not ripen appreciably after picking, most fruits are picked as near to eating-ripe maturity as possible.

Pre-process Handling and Operations

Freezing preservation of fruits can only help retain the inherent quality present initially in a product since the process does not improve the quality characteristics of raw materials. Therefore, quality level of the raw materials prior to freezing is the major consideration for successful freezing. Washing and cutting generally results in losses when applied after thawing. Thus, fruits should be prepared prior to the freezing process in terms of peeling, slicing or cutting. Freezing preservation does not require specific unit operations for cleaning, rinsing, sorting, peeling, and cutting of fruits.

Fruits that require peeling before consumption should be peeled prior to freezing. Peeling is done by scalding the fruit in hot water, steam or hot lye solutions (Boyle and Wolford, 1968). The effect of peeling on the quality of frozen products has been studied for several fruits, including kiwi, banana, and mang. The rate of freezing can be increased by decreasing the size of products frozen, especially for large fruits. An increase in the freezing rate results in smaller ice crystals, which decreases cellular damage in fruit tissue. Banana, tomato, mango, and kiwi are some examples of large fruits commonly cut into smaller cubes or slices prior to freezing.

The objective of blanching is to inactivate the enzymes causing detrimental changes in colour, odour, flavour, and nutritive value, but heat treatment causes loss of such characteristics in fruits. Therefore, only a few types of fruits are blanched for inactivation of enzymes prior to freezing. The loss of water-soluble minerals and vitamins during blanching should also be minimized by keeping blanching time and temperature at an optimum combination.

Effect of Ingredients

Addition of sugars is an extremely important pretreatment for fruits prior to freezing since the treatment has the effect of excluding oxygen from the fruit, which helps to retain colour and appearance. Sugars when dissolved in solutions act by withdrawing water from cells by osmosis, resulting in very concentrated solutions inside the cells. The high concentration of solutes depresses the freezing point and therefore reduces the freezing within the cells, which inhibits excessive structural damage. Sugar syrups in the range of 30-60 percent sugar content are commonly used to cover the fruit completely, acting as a barrier to oxygen transmission and browning. Several experiments have shown the protective effect of sugar on flavour, odour, colour, and nutritive value during freezing, especially for frozen berries.

Packaging

Fruits exposed to oxygen are susceptible to oxidative degradation, resulting in browning and reduced storage life of products. Therefore, packaging of frozen fruits is based on excluding air from the fruit tissue. Replacement of oxygen with sugar solution or inert gas, consuming the oxygen by glucose-oxidase and/or the use of vacuum and oxygen-impermeable films are some of the methods currently employed for packaging frozen fruits. Plastic bags, plastic pots, paper bags, and cans are some of the most commonly used packaging materials (with or without oxygen removal) selected, based on penetration properties and thickness.

There are several types of fruit packs suitable for freezing: syrup pack, sugar pack, unsweetened pack, and tray pack and sugar replacement pack. The type of pack is usually selected according to the intended use for the fruit. Syrup-packed fruits are generally used for cooking purposes, while dry-packed and tray-packed fruits are good for serving raw in salads and garnishes.

Syrup Pack

The proportion of sugar to water used in a syrup pack depends on the sweetness of the fruit and the taste preference of the consumer. For most fruits, 40 percent sugar syrup is recommended. Lighter syrups are lower in calories and mostly desirable for mild-flavoured fruits to prevent masking the flavour, while heavier syrups may be used for very sour fruits.

Syrup is prepared by dissolving the sugar in warm water and cooling the solution down before usage. Just enough cooled syrup is used to cover the prepared fruit after it has been settled by jarring the container. In order to keep the fruit under the syrup, a small piece of crumpled waxed paper or other water resistant wrapping material is placed on top; the fruit is pressed down into the syrup before closing, then sealed and frozen.

Pectin can be used to reduce sugar content in syrups when freezing berries, cherries, and peaches. Pectin syrups are prepared by dissolving 1 box of powdered pectin with 1 cup of water. The solution is stirred and boiled for 1 minute; 1/2 cup of sugar is added and dissolved; the solution is then cooled down with the addition of cold water. Previously prepared fruit is put into a 4 to 6 quart bowl and enough pectin syrup is added to cover the fruit with a thin film. The pack is sealed and promptly frozen.

Sugar Packs

In preparing a sugar pack, sugar is first sprinkled over the fruit. Then the container is agitated gently until the juice is drawn out and the sugar is dissolved. This type of pack is generally used for soft sliced fruits such as peaches, strawberries, plums, and cherries, by using sufficient syrup to cover the fruit. Some whole fruits may also be coated with sugar prior to freezing.

Unsweetened Packs

Unsweetened packs can be prepared in several ways, either dry-packed, covered with water containing ascorbic acid, or packed in unsweetened juice. When water or juice is used in syrup and sugar packs, fruit is submerged by using a small piece of crumpled water-resistant material. Generally, unsweetened packs yield a lower quality product when compared with sugar packs, with the exception, some fruits such as raspberries, blueberries, scalded apples, gooseberries, currants, and cranberries maintain good quality without sugar.

Tray Packs

Unsweetened packs are generally prepared by using tray packs in which a single layer of prepared fruit is spread on shallow trays, frozen, and packaged in freezer bags promptly. The fruit sections remain loose without clumping together, which offers the advantage of using frozen fruit piece by piece.

Sugar Replacement Packs

Artificial sweeteners can be used instead of sugar in the form of sugar substitutes. The sweet taste of sugar can be replaced by using these kinds of sweeteners, however the beneficial effects of sugar like colour protection and thick syrup can not be replaced. Fruits frozen with sugar substitutes will freeze harder and thaw more slowly than fruits preserved with sugar.

Freezing Vegetables

Freezing is often considered the simplest and most natural way of preservation for vegetables. Frozen vegetables and potatoes form a significant proportion of the market in terms of frozen food consumption. The quality of frozen vegetables depends on the quality of fresh products, since freezing does not improve product quality. Pre-process handling, from the time vegetables are picked until ready to eat, is one of the major concerns in quality retention.

Fruit	Preparation	Type of Pack
Apples	Wash, peel, and slice into antidarkening solution - 3 tablespoons lemon juice per quart of water	Pack in 30-40% syrup, adding 1/2 teaspoon crystalline ascorbic acid per quart of syrup. Pack dry or with up to 1/2 cup sugar per quart of apple slices.
Apricots	Wash, halve, and pit. Peel and slice if desired. If apricots are not peeled, heat in boiling water for 1/2 minute to keep skins from toughening during freezing. Cool in cold water, drain.	Pack in 40% syrup, adding 3/4 teaspoon crystalline ascorbic acid per quart of syrup.
Avocados	Peel soft, ripe avocados. Cut in half, remove pit, mash pulp.	Add 1/8 teaspoon crystalline ascorbic acid to each quart of puree. Package in recipe-size amounts.
Berries	Select firm, fully ripe berries. Sort, wash, and drain.	Use 30% syrup pack, dry unsweetened pack, dry sugar pack, (3/4 cup sugar per quart of berries), or tray pack.

Cherries(sour or sweet)	Select well-colored, tree-ripened cherries. Stem, sort, and wash thoroughly. Drain and pit.	Pack in 30-40% syrup. Add 1/2 teaspoon ascorbic acid per quart of syrup. For pies and other cooked products, pack in dry sugar using 3/4-cup sugar per quart of fruit.
Citrus fruits, (sections or slices)	Select firm fruit, free of soft spots. Wash and peel.	Pack in 40% syrup or in fruit juice. Add 1/2 teaspoon ascorbic acid per quart of syrup or juice.
Grapes	Select firm, ripe grapes. Wash and remove stems. Leave seedless grapes whole. Cut grapes with seeds in half and remove seeds.	Pack in 20% syrup or pack without sugar. Use dry pack for halved grapes and tray pack for whole grapes.
Melons (cantaloupe, watermelon)	Select firm-fleshed, well-colored, ripe melons. Wash rinds well. Slice or cut into chunks.	Pack in 30% syrup or pack dry using no sugar. Pulp also may be crushed (except watermelon), adding 1 tablespoon sugar per quart. Freeze in recipe-size containers.

Crop Cultivar, Production and Maturity

The choice of the right cultivar and maturity before crop is harvested are the two most important factors affecting raw material quality. Raw material characteristics are usually related to the vegetable cultivar, crop production, crop maturity, harvesting practices, crop storage, transport, and factory reception.

The choice of crop cultivars is mostly based on their suitability for frozen preservation in terms of factory yield and product quality. Some of the characteristics used as selection criteria are as follows:

- Suitability for mechanical harvesting
- Uniform maturity
- Exceptional flavour and uniform colour and desirable texture
- Resistance to diseases
- High yield

Although cultivar selection is a major factor affecting the quality of the final product, many practices in the field and factors during growth of crop can also have a significant effect on quality. Those practices include site selection for growth, nutrition of crop, and use of agricultural chemicals to control pests or diseases. The maturity assessment for harvesting is one of the most difficult parts of the production. In addition to conventional methods, new instruments and tests have been developed to predict the maturity of crops that help determining the optimum harvest time, although the maturity assessment differs according to crop variety.

Harvesting

At optimum maturity, physiological changes in several vegetables take place very rapidly. Thus, the determination of optimum harvesting time is critical. Some vegetables such as green peas and sweet corn only have a short period during which they are of prime quality. If harvesting is delayed beyond this point, quality deteriorates and the crop may quickly become unacceptable. Most of the vegetables are subjected to bruising during harvesting.

Pre-process Handling

Vegetables at peak flavour and texture are used for freezing. Postharvest delays in handling vegetables are known to produce deterioration in flavour, texture, colour, and nutrients. Therefore, the delays between harvest and processing should be reduced to retain fresh quality prior to freezing. Cooling vegetables by cold water, air blasting, or ice will often reduce the rate of post-harvest losses sufficiently, providing extra hours of high quality retention for transporting raw material to considerable distances from the field to the processing plant.

Blanching

Blanching is the exposure of the vegetables to boiling water or steam for a brief period of time to inactivate enzymes. Practically every vegetable (except herbs and green peppers) needs to be blanched and promptly cooled prior to freezing, since heating slows or stops the enzyme action, which causes vegetables to grow and mature. After maturation, however, enzymes can cause loss in quality, flavour, colour, texture, and nutrients. If vegetables are not heated sufficiently, the enzymes will continue to be active during frozen storage and may cause the vegetables to toughen or develop off-flavours and colours. Blanching also causes wilting or softening of vegetables, making them easier to pack. It destroys some bacteria and helps remove any surface dirt.

Blanching in hot water at 70 to 105°C has been associated with the destruction of enzyme activity. Blanching is usually carried out between 75 and 95°C for 1 to 10 minutes, depending on the size of individual vegetable pieces. Blanched vegetables should be promptly cooled down to control and minimize the degradation of soluble and heat-labile nutrients.

The enzymes used as indicators of effectiveness of the blanching treatment are peroxidase, catalase, and more recently lipoxygenase. Peroxidase inactivation is commonly used in vegetable processing, since peroxidase is easily detected and is the most heat stable of these enzymes.

Vegetables can be blanched in hot water, steam, and in the microwave. Hot water blanching is the most common way of processing vegetables. Blanching times recommended for various vegetables are given in table, which indicates that the operation time can vary depending on the intended product use. For water blanching, vegetables are put in a basket and then placed in a kettle of boiling water covered with a lid. Timing begins immediately. Steam blanching takes longer than the water method, but helps retain water-soluble nutrients such as water-soluble vitamins. For steam blanching, a single layer of vegetables is placed on a rack or in a basket at 3-5 cm above water boiling in a kettle. A tightly fitted lid is placed on the kettle and timing is started. Microwave blanching is usually recommended for small quantities of vegetables prior to freezing. Due to the non-uniform heating disadvantage of microwaves, research is still being conducted to obtain better results with microwave blanching.

Vegetable	Preparation	Blanch/Freeze	
Asparagus	Wash and sort by sizeSnap off tough ends. Cut stalks into 5-cm lengths.	Water blanch:	2 min
		Steam blanch:	3 min
Beans	Wash and trim the ends. Cut if desired.	Water blanch:	Steam blanch:
		Whole: 3 min.	Whole: 4 min.
		Cut: 2min.	Cut: 3min.

Beets	Wash and remove the tops leaving 2.5 cm of stem and root.	Cook until tender: 25-30 min Cool promptly, peel, trim. Cut into slices or cubes and pack.	
Broccoli	Wash and cut into pieces.	Water blanch:	3 min.
		Steam blanch:	3 min.
Cabbage	Wash and cut into wedges.	Water blanch:	3 min.
		Steam blanch:	4 min.
Carrots	Wash, peel and trim. Cut if desired.	Water blanch: 5 min.	
Cauliflower	Discard leaves; steam and wash. Break into flowerets.	Water blanch:	Steam blanch:
		Whole: 5 min.	Whole: 7 min
Corn	Remove husks and silks. Trim ends and wash.	Water blanch:	Steam blanch:
		Whole: 5 min.	Whole: 7 min
Greens	Select young tender greens. Wash and trim the leaves.	Water blanch:	2 min.
		Steam blanch:	3 min.
Herbs	Wash.	No heat treatment is needed.	
Mushrooms	Wipe and damp with paper towel. Trim hard tip of stems. Sort and cut large mushrooms.	May be frozen without heat treatment.	
Peas	Shell garden peas.	Water blanch:	Steam blanch:
		1-1/2 min.	1-1/2 min.
Peppers	Wash, remove stems and seeds.	Freeze whole or cut as desired. No heat treatment is needed.	
Potatoes	Peel, cut or grate as desired.	Water blanch:	
		Whole: 5 min.	
		Pieces: 2-3 min.	

Packaging

There are several factors to consider in packaging frozen vegetables, which include protection from atmospheric oxygen, prevention of moisture loss, retention of flavour, and rate of heat transfer through the package . There are two basic packing methods recommended for frozen vegetables: dry pack and tray pack.

In the dry pack method, the blanched and drained vegetables are put into meal-sized freezer bags and packed tightly to cut down on the amount of air in the package. Proper headspace (approximately 2 cm) is left at the top of rigid containers before closing. For freezer bags, the headspace is larger. Provision for headspace is not necessary for foods such as broccoli, asparagus, and brussels sprouts, as they do not pack tightly in containers.

In the tray pack method, chilled, well-drained vegetables are placed in a single layer on shallow trays or pans. Trays are placed in a freezer until the vegetables become firm, then removed. Vegetables are filled into containers. Tray-packed foods do not freeze in a block but remain loosely distributed so that the amount needed can be poured from the container and the package reclosed.

Physical Aspects of Freezing

Moisture migration is the principal physical change occurring in frozen foods, affecting the physical, chemical, and biochemical properties, including texture and palatability of the food.

Texture

Most fruits and vegetables are over 90 percent water of total weight. The water and dissolved solutes inside the rigid plant cell walls give support to the plant structure, and texture to the fruit or vegetable tissue. In the process of freezing, when water in the cells freezes, an expansion occurs and ice crystals cause the cell walls to rupture. Consequently, the texture of the produce is generally much softer after thawing when compared to non-frozen produce. This textural difference is especially noticeable in products normally consumed raw, as in the case of fruits. It is usually recommended that frozen fruits be served before they are completely thawed, since in the partially thawed state the effect of freezing on the fruit tissue is less noticeable. On the other hand, due to the fact cooking also softens cell walls, textural changes caused by freezing are not significantly noticeable in products cooked before eating, as in the case of most vegetables.

Freezer Burn

One of the most common forms of quality degradation due to moisture migration in frozen foods is freezer burn, a condition defined as the glassy appearance in some frozen products produced by ice crystals evaporating on the surface area of a product. The grainy, brownish spots occurring on the product cause the tissue to become dry and tough and to develop off-flavors. This quality defect can be prevented by using heavyweight, moisture proof packaging during the freezing process.

Chemical Aspects of Freezing

Chemical changes that can cause spoilage and deterioration of fresh fruits and vegetables will continue after harvesting. This is the main reason for eliminating any delays during pre-freezing operations. As mentioned earlier, several enzymes that cause the loss of color, loss of nutrients, flavor changes, and color changes in frozen fruits and vegetables, should be inactivated by means of thermal treatments prior to freezing. In most cases, blanching is essential for producing quality frozen vegetables, since it also helps destroy microorganisms on the surface of the produce. However, in processing fruits, heat treatment may cause more degradation in quality. In this case, enzymes in frozen fruits can be controlled by using chemical compounds, which interfere with deteriorative chemical reactions. Ascorbic acid is an example that may be used in its pure form or in commercial mixtures with sugars for inhibition of enzymes in fruits.

Development of rancid oxidative flavors through contact of the frozen product with air is another group of chemical changes that can take place in frozen products. This problem can be controlled by excluding oxygen through proper packaging as mentioned earlier. It is also advisable to remove as much air as possible from the freezer bag or container to reduce the amount of air in contact with the product.

Formulation and Processing of Selected Frozen Food Prototypes

Selecting a Formulation for Mixed Fruits

One of the most common commercially frozen fruit products is mixed frozen berries. The mixtures typically contain combinations of raspberries, blackberries, blueberries, and strawberries. There is a wide range of mixtures available in the market for frozen berry combinations. Raspberries and blackberries for example, which are known to freeze well and retain their wholeness and shape, dependent on the structure of the fruit, are strongly associated with their cultivar. The processing requirements for different varieties of berries do not change significantly. Therefore, a mixture of raspberries and blackberries is chosen in this case as a fruit formulation to simplify the freezing process.

Raspberries and blackberries.

Procedure for processing mixed berries

- Full-flavoured, ripe berries of like size preferably with tender skins are selected;

- Berries are sorted, washed, and drained;

- Berries are packed into containers and covered with cold 40 percent sugar syrup, with proper headspace;

- Polyethylene freezer bags are sealed and frozen.

Selecting a Formulation for Mixed Vegetables

Mixed vegetables

Frozen mixed vegetables constitute a large portion of the frozen vegetable market and are now available in an ever-increasing variety of mixtures. The mixtures include three or more types of vegetables, properly prepared and blanched. The USDA standards for frozen mixed vegetables describe this item as a mixture containing three or more of the basic vegetables - beans, carrots, corn, and peas. When three vegetables are used, none of the vegetables should be more than 40 percent of the total weight; the individual percent decreases with increased number of vegetable types.

In a mixed frozen vegetable product, vegetables of different sizes are present in the mixture. Therefore, during pre-freeze treatments, especially blanching, care must be taken to be sure all vegetables are blanched properly.

Procedure for processing mixed vegetables

- A mixture of four vegetables in which none of the vegetables is less than 8 percent by weight nor more than 35 percent by weight of all the frozen mixed vegetables are selected;

- The vegetables are sorted, washed and peeled;

- They are cut into uniform size and blanched in hot water for 5 minutes, and immediately cooled after blanching;

- Packed and frozen.

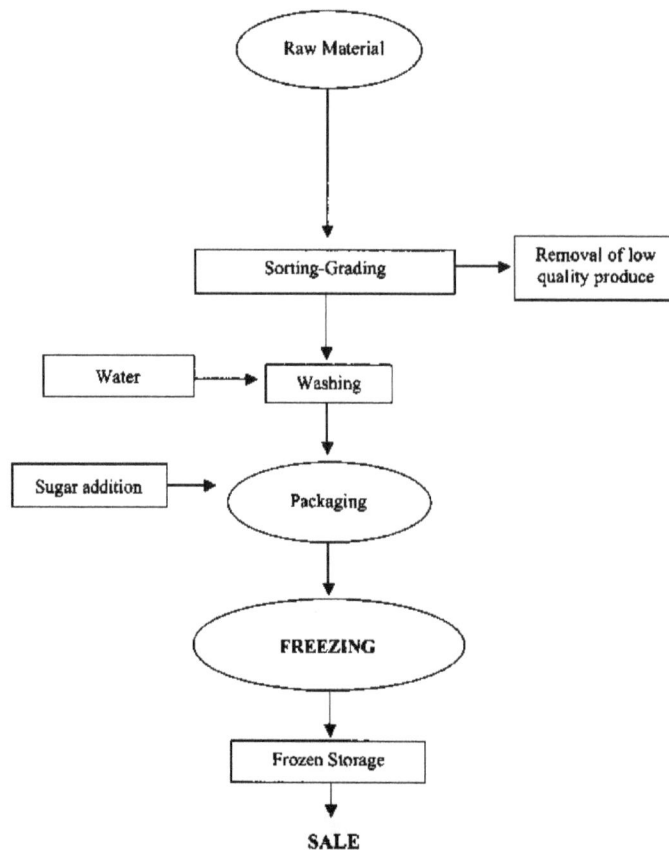

Flow diagram of freezing process for fruit-based product.

Boiling

Boiling is the process of applying heat to water until the temperature reaches about 100°C. Boiling foods in water cannot completely destroy all microorganisms, but the vegetative cells of bacteria, yeasts and moulds are generally quickly destroyed at temperatures of 100°C or above. Spores of some bacteria are extremely resistant to heat and are not killed at this temperature, although their growth is prevented. For this reason, boiling food can rarely be relied upon to ensure complete destruction of all organisms. However, most pathogens are killed, provided that sufficient exposure time is maintained. Although the spores of Clostridium botulinum, which causes botulism, are extremely heat-resistant, the toxin produced by this organism is readily destroyed by boiling. However, some toxins produced by other bacteria such as staphylococci are not easily inactivated. Thermophilic (heat-loving) organisms may survive the effects of boiling and can cause food spoilage if environmental conditions are favourable for them.

Types

Nucleate

Nucleate boiling is characterized by the growth of bubbles or pops on a heated surface, which rises from discrete points on a surface, whose temperature is only slightly above the liquids. In general, the number of nucleation sites are increased by an increasing surface temperature.

Nucleate boiling of water over a kitchen stove burner

An irregular surface of the boiling vessel (i.e., increased surface roughness) or additives to the fluid (i.e., surfactants and/or nanoparticles) can create additional nucleation sites, while an exceptionally smooth surface, such as plastic, lends itself to superheating. Under these conditions, a heated liquid may show boiling delay and the temperature may go somewhat above the boiling point without boiling.

Critical Heat Flux

As the boiling surface is heated above a critical temperature, a film of vapor forms on the surface. Since this vapor film is much less capable of carrying heat away from the surface, the temperature rises very rapidly beyond this point into the transition boiling regime. The point at which this occurs is dependent on the characteristics of boiling fluid and the heating surface in question.

Transition

Transition boiling may be defined as the unstable boiling, which occurs at surface temperatures between the maximum attainable in nucleate and the minimum attainable in film boiling.

The formation of bubbles in a heated liquid is a complex physical process which often involves cavitation and acoustic effects, such as the broad-spectrum hiss one hears in a kettle not yet heated to the point where bubbles boil to the surface.

Film

If a surface heating the liquid is significantly hotter than the liquid then film boiling will occur, where a thin layer of vapor, which has low thermal conductivity, insulates the surface. This condition of a vapor film insulating the surface from the liquid characterizes *film boiling*.

Uses

In Cooking

Boiling is the method of cooking food in boiling water or other water-based liquids such as stock or milk. Simmering is gentle boiling, while in poaching the cooking liquid moves but scarcely bubbles.

Boiling pasta

The boiling point of water is typically considered to be 100°C or 212°F. Pressure and a change in the composition of the liquid may alter the boiling point of the liquid. For this reason, high elevation cooking generally takes longer since boiling point is a function of atmospheric pressure. In Denver, Colorado, USA, which is at an elevation of about one mile, water boils at approximately 95°C or 203°F. Depending on the type of food and the elevation, the boiling water may not be hot enough to cook the food properly. Similarly, increasing the pressure as in a pressure cooker raises the temperature of the contents above the open air boiling point.

Some science suggests adding a water-soluble substance, such as salt or sugar also increases the boiling point. This is called boiling-point elevation. At palatable concentrations of salt, the effect is very small, and the boiling point elevation is difficult to notice and this is why experiments to prove this are considered inconclusive. However, while making thick sugar syrup, such as for Gulab Jamun, one will notice boiling point elevation. Due to variations in composition and pressure, the boiling point of water is almost never exactly 100 °C, but rather close enough for cooking.

Boiling milk

Foods suitable for boiling include vegetables, starchy foods such as rice, noodles and potatoes, eggs, meats, sauces, stocks, and soups.

Boiling has several advantages. It is safe and simple, and it is appropriate for large-scale cookery. Older, tougher, cheaper cuts of meat and poultry can be made digestible. Nutritious, well-flavored stock is produced. Also, maximum color and nutritive value is retained when cooking green vegetables, provided boiling time is kept to the minimum.

Boiling food over a fireplace

On the other hand, there are several disadvantages. There is a loss of soluble vitamins from foods to the water (if the water is discarded). Boiling can also be a slow method of cooking food.

Boiling can be done in several ways: The food can be placed into already rapidly boiling water and left to cook, the heat can be turned down and the food can be simmered or the food can also be placed into the pot, and cold water may be added to the pot. This may then be boiled until the food is satisfactory.

Water on the outside of a pot, i.e., a wet pot, increases the time it takes the pot of water to boil. The pot will heat at a normal rate once all excess water on the outside of the pot evaporates.

Boiling is also often used to remove salt from certain foodstuffs, such as bacon if a less saline product is required.

Boil-in-the-bag

Also known as "boil-in-bag", this involves heating or cooking ready-made foods sealed in a thick plastic bag. The bag containing the food, often frozen, is submerged in boiling water for a prescribed time. The resulting dishes can be prepared with greater convenience as no pots or pans are dirtied in the process. Such meals are available for camping as well as home dining.

Heating

Heat treatment or thermal processing of food is used in order to kill or inactivate bacteria, increase the shelf-life, or create products with an attractive appearance. Today's cooled displays contain healthy and palatable ready-to-eat meals several weeks after they were produced. This is possible because the food has undergone heat treatment during production.

Minimal Thermal Processing

The food is heated to a low temperature for a short period and acquires a fresh and attractive appearance, while retaining most of the nutrients. The technique is particularly suitable for food that does not need a shelf-life of longer than 7-12 days. Catering establishments for canteens, sheltered accommodation and hospitals often use this technique, and it is used also by restaurants and other food outlets.

Cook-chill

The cook-chill method is used to preserve as much flavour and moisture as possible in the products. The food is heat treated at a temperature and time depending on raw material and desired product quality. After rapid cooling the food is packaged and stored at chilled temperatures. The shelf-life of such products is generally limited to 10-14 days, depending on the storage conditions and raw material. The method is often combined with modified atmosphere packaging.

Sous Vide

Sous vide is used to describe food that has been vacuum-packed and given mild heat treatment before being stored in at chilled conditions. The heat treatment takes place after packaging and under controlled conditions of time and temperature, and the food is subsequently rapidly cooled. The products are kept chilled until they are heated before serving. Sous vide is now used with a more intense heat treatment, and this has made it possible to produce ready-to-eat food that has a shelf-life of several weeks. Testing the nutritional value of four ready-to-eat meals has shown the method to be just as good as corresponding home-cooking.

Pickling

The term pickle is derived from the Dutch word pekel, meaning brine. Pickling, also known as brining or corning is the process of preserving food by anaerobic fermentation in brine (a solution of salt in water) to produce lactic acid, or marinating and storing it in an acid solution, usually vinegar (acetic acid). The resulting food is called a pickle. This procedure gives the food a salty or sour taste. In South Asia, edible oils are used as the pickling medium with vinegar. Another distinguishing characteristic is a pH less than 4.6, which is sufficient to kill most bacteria. Pickling can preserve perishable foods for months. Antimicrobial herbs and spices, such as mustard seed, garlic, cinnamon or cloves, are often added. If the food contains sufficient moisture, a pickling brine may be produced simply by adding dry salt. For example, sauerkraut and Korean kimchi are produced by salting the vegetables to draw out excess water. Natural fermentation at room temperature, by lactic acid bacteria, produces the required acidity. Other pickles are made by placing vegetables in vinegar. Unlike the canning process, pickling (which includes fermentation) does not require that the food be completely sterile before it is sealed. The acidity or salinity of the solution, the temperature of fermentation, and the exclusion of oxygen determine which microorganisms dominate, and determine the flavor of the end product When both salt concentration and temperature are low, Leuconostocmesenteroides dominates, producing a mix of acids, alcohol, and aroma compounds. At higher temperatures Lactobacillus plantarum dominates, which produces primarily lactic acid. Many pickles start with Leuconostoc, and change to Lactobacillus with higher acidity. Pickling began as a way to preserve food for out-of-season use and for long journeys, especially by sea. Salt pork and salt beef were common staples for sailors before the days of steam engines. Although the process was invented to preserve foods, pickles are also made and eaten because people enjoy the resulting flavors. Pickling may also improve the nutritious value of food by introducing B vitamins produced by bacteria.

The exact origins of pickling are unknown, but the ancient Mesopotamians may have used the process around 2400 B.C. Pickling was used as a way to preserve food for out-of-season use and for long journeys, especially by sea. Salt pork and salt beef were common staples for sailors before the days of steam engines. Although the process was invented to preserve foods, pickles are also made and eaten because people enjoy the resulting flavors. Pickling may also improve the nutritional value of food by introducing B vitamins produced by bacteria.

The term pickle is derived from the Dutch word pekel, meaning brine. In the U.S. and Canada, and sometimes Australia and New Zealand, the word pickle alone almost always refers to a pickled cucumber, except when it is used figuratively. It may also refer to other types of pickles such as "pickled onion", "pickled cauliflower", etc. In the UK, pickle, as in a "cheese and pickle sandwich", may also refer to Ploughman's pickle, a kind of chutney.

Process of Pickling

Pickling is a method of preserving food in an edible anti-microbial liquid. Pickling is of two types:

- Chemical pickling (brining)
- Fermentation pickling

In chemical pickling, food is placed in an edible liquid that inhibits or kills bacteria and other micro-organisms. A number of pickling agents can be used like brine, vinegar, alcohol, and vegetable oil (olive or mustard oil). The chemical pickling process could also involve heating or boiling so that the food being preserved becomes saturated with the pickling agent. Common chemically pickled foods include cucumbers, peppers, corned beef, herring, and eggs, chow-chow (chayote and Hindi *chocho)* as well mixed vegetables such as piccalilli and giardiniera (chopped pickled vegetables and spices) and also Indian achar.

In fermented pickling, the food itself produces the preservation agent, typically by a process that produces lactic acid. Fermented pickles include sauerkraut, nukazuke (Japanese pickle made by fermenting vegetables in rice bran) kimchi, surströmming (Swedish food- fermented Baltic Sea herring) and curtido (fermented cabbage, onions, carrots, oregano- Central American cuisine). Some chemically pickled cucumbers are also fermented. In commercial pickles, a preservative like sodium benzoate or EDTA may also be added to increase shelf life.

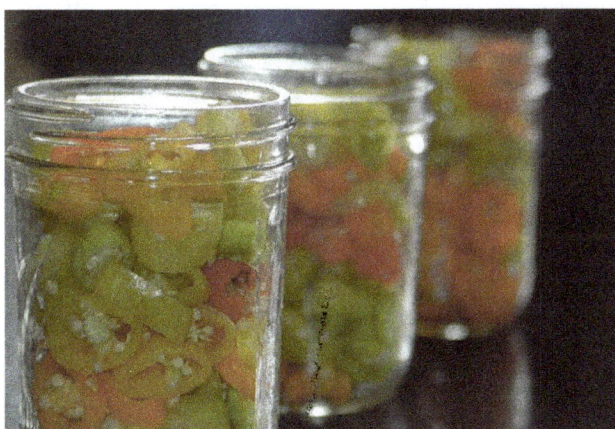

Popularity of Pickles Around the World

South Asia

South Asia has a large variety of pickles (known as *achar* in Assamese, Bengali, Hindi, Punjabi, *uppinakaayi* in Kannada, *lonacha* in Marathi, *uppilittathu* or *achar* in Malayalam, *oorukai* in Tamil, *ooragaya* in Telugu), which are mainly made from varieties of mango, lemon, lime, goongura(a sour leafy shrub), tamarind and Indian gooseberry (amla), chilli. Vegetables such as eggplant, carrots, cauliflower, tomato, bitter gourd, green tamarind, ginger, garlic, onion, and citron are also occasionally used. These fruits and vegetables are generally mixed with ingredients like salt, spices, and vegetable oils and are set to mature in a moistureless medium.

In Pakistan, pickles are known locally as *achaar* (in Urdu) and come in a variety of flavors. A popular item is the traditional mixed Hyderabadi pickle, a common delicacy prepared from an assortment of fruits (most notably mangoes) and vegetables blended with selected spices.

In Sri Lanka, *achcharu* is traditionally prepared from carrots, onions, and ground dates that are mixed with mustard powder, ground pepper, crushed ginger, garlic, and vinegar, and left to sit in a clay pot.

Southeast Asia

Singapore, Indonesian and Malaysian pickles, called *acar*, are typically made out of cucumber, carrot, bird's eye chilies, and shallots, these items being seasoned with vinegar, sugar and salt. Fruits, such as papaya and pineapple, are also sometimes pickled.

In the Philippines, *achara* is primarily made out of green papaya, carrots, and shallots, with cloves of garlic and vinegar. Other versions could include ginger, bell peppers, white radishes, cucumbers or bamboo shoots. Separately, in some provinces, unripe mangoes or *burong mangga*, unripe tomatoes, guavas, jicama, bitter gourd and other fruit and vegetables are also pickled. Siling labuyo, sometimes with garlic and red onions, are also pickled in bottled vinegar. The spiced vinegar itself is a staple condiment in Filipino cuisine.

In Vietnam, vegetable pickles are called *dưa muối* ("salted vegetables") or *dưa chua* ("sour vegetables"). In Burma, tea leaves are pickled to produce lahpet, which has strong social and cultural importance.

Kimchi is a very common side dish in Korea.

East Asia

China is home to a huge variety of pickled vegetables, including radish, *baicai* (Chinese cabbage, notably *suan cai*, *la bai cai*, *pao cai*, and Tianjin preserved vegetable), *zha cai*, chili pepper, and cucumbers, among many others.

Japanese *tsukemono* (pickled foods) include *takuan* (daikon), *umeboshi* (ume plum), *gari* & *beni shoga* (ginger), turnip, cucumber, and Chinese cabbage.

The Korean staple kimchi is usually made from pickled napa cabbage and radish, but is also made from green onions, garlic stems, chives and a host of other vegetables. Kimchi is popular throughout East Asia. Jangajji is another example of pickled vegetables.

Western Asia

Torshi, traditional pickles in Southeast Europe, Western Asia and the Caucasus

In Iran, Turkey, Arab countries, the Balkans, and the Caucasus, pickles (called torshi in Persian, *turşu* in Turkish language and *mekhallel* in Arabic) are commonly made from turnips, peppers, carrots, green olives, cucumbers, cabbage, green tomatoes, lemons, and cauliflower.

Central and Eastern Europe

Coriander seeds are one of the spices popularly added to pickled vegetables in Europe.

In Hungary the main meal *(lunch)* usually goes with some kind of pickles *(savanyúság)* but they are commonly consumed at other times of the day too. The most commonly consumed pickles are sauerkraut *(savanyú káposzta)*, the different kinds of pickled cucumbers and peppers and *csalamádé* but tomatoes, carrots, beetroot, baby corn, onions, garlic, certain squashes and melons and a few fruits like plums and apples are used to make pickles too. Stuffed pickles are specialties usually made of peppers or melons pickled after being stuffed with a cabbage filling. Pickled plum stuffed with garlic is a unique

Hungarian type of pickle just like *csalamádé* and leavened cucumber *(kovászos uborka)*. *Csalamádé* a type of mixed pickle made of cabbage, cucumber, paprika, onion, carrot, tomatoes and bay leaf mixed up with vinegar as the fermenting agent. Leavened cucumber, unlike other types of pickled cucumbers that are around all year long, is rather a seasonal pickle produced in the summer. Cucumbers, spices, herbs and slices of bread are put in a glass jar with salt water and kept in direct sunlight for a few days. The yeast from the bread, along with other pickling agents and spices fermented under the hot sun, give the cucumbers a unique flavor, texture and slight carbonation. Its juice can be used to make a special type of spritzer *('Újházy fröccs')* instead of carbonated water. It is common for Hungarian households to produce their own pickles. Different regions or towns have their special recipes unique to them. Among them all the Vecsési Sauerkraut *(Vecsési savanyú káposzta)* is the most famous.

Pickled tomatoes are common in CIS

Romanian pickles (murături) are made out of beetroot, cucumbers, green tomatoes *(gogonele)*, carrots, cabbage, garlic, sauerkraut (bell peppers stuffed with cabbage), bell peppers, melons, mushrooms, turnips, celery and cauliflower. Meat, like pork, can also be preserved in salt and lard.

Polish, Czech and Slovak traditional pickles are cucumbers and sauerkraut, but other pickled fruits and vegetables, including plums, pumpkins and mushrooms are also common.

Russian, Ukrainian and Belarusian pickled items include beets, mushrooms, tomatoes, sauerkraut, cucumbers, ramsons, garlic, eggplant (which is typically stuffed with julienned carrots), custard squash, and watermelon. Garden produce is commonly pickled using salt, dill, blackcurrant leaves, bay leaves and garlic and is stored in a cool, dark place. The leftover brine (called *rassol* (рассол) in Russian) has a number of culinary uses in these countries, especially for cooking traditional soups, such as shchi, rassolnik, and solyanka. *Rassol*, especially cucumber or sauerkraut rassol, is also a favorite traditional remedy against morning hangover.

Southern Europe

An Italian pickled vegetable dish is giardiniera, which includes onions, carrots, celery and cauliflower. Many places in southern Italy, particularly in Sicily, pickle eggplants and hot peppers.

In Albania, Bulgaria, Serbia, Macedonia and Turkey, mixed pickles, known as *turshi*, *tursija* or *turshu* form popular appetizers, which are typically eaten with *rakia*. Pickled green tomatoes, cucumbers, carrots, bell peppers, peppers, eggplants, and sauerkraut are also popular.

Turkish pickles, called *turşu*, are made out of vegetables, roots, and fruits such as peppers, cucumber, Armenian cucumber, cabbage, tomato, eggplant (aubergine), carrot, turnip, beetroot, green almond, baby watermelon, baby cantaloupe, garlic, cauliflower, bean and green plum. A mixture of spices flavor the pickles.

In Greece, pickles, called *τουρσί(α),* are made out of carrots, celery, eggplants stuffed with diced carrots, cauliflower, tomatoes, and peppers.

Northern Europe

In Britain, pickled onions and pickled eggs are often sold in pubs and fish and chip shops. Pickled beetroot, walnuts, and gherkins, and condiments such as Branston Pickle and piccalilli are typically eaten as an accompaniment to pork pies and cold meats, sandwiches or a ploughman's lunch. Other popular pickles in the UK are pickled mussels, cockles, red cabbage, mango chutney, sauerkraut, and olives. Rollmops are also quite widely available under a range of names from various producers both within and out of the UK.

Pickled herring, rollmops, and salmon are popular in Scandinavia. Pickled cucumbers and red garden beets are important as condiments for several traditional dishes. Pickled capers are also common in Scandinavian cuisine.

United States and Canada

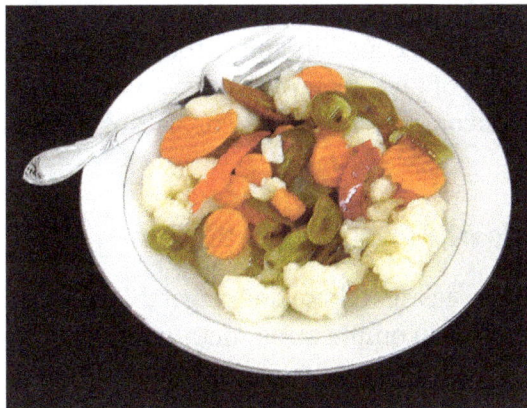

A dish of giardiniera

In the United States and Canada, pickled cucumbers (most often referred to simply as "pickles" in Canada and the United States), olives, and sauerkraut are most commonly seen, although pickles common in other nations are also available.

Canadian pickling is similar to that of Britain. Through the winter, pickling is an important method of food preservation. Pickled cucumbers, onions, and eggs are common individual pickled foods seen in Canada. Chow-chow is a tart vegetable mix popular in the Maritime Provinces and the Southern United States, similar to piccalilli. Pickled fish is commonly seen, as in Scotland. Meat is often also pickled or preserved in different brines throughout the winter, most prominently in the harsh climate of Newfoundland.

Giardiniera, a mixture of pickled peppers, celery and olives, is a popular condiment in Chicago and other cities with large Italian-American populations, and is often consumed with Italian beef sandwiches. Pickled eggs are common in the Upper Peninsula of Michigan. Pickled herring is available in the Upper Midwest. Pennsylvania Dutch Country has a strong tradition of pickled foods, including chow-chow and red beet eggs. In the Southern United States, pickled okra and watermelon rind are popular, as are deep-fried pickles and pickled pig's feet, pickled chicken eggs, pickled quail eggs, pickled garden vegetables and pickled sausage. In Mexico, chili peppers, particularly of the Jalapeño and serrano varieties, pickled with onions, carrots and herbs form common condiments. Various pickled vegetables, fish, or eggs may make a side dish to a Canadian lunch or dinner. Popular pickles in the Pacific Northwest include pickled asparagus and green beans. Pickled fruits like blueberries and early green strawberries are paired with meat dishes in restaurants.

In the United States, National Pickle Day is recognized as a food "holiday" every year on November 14.

Mexico, Central America and South America

In the Mesoamerican region pickling is known as "encurtido" or "curtido" for short. The pickles or "curtidos" as known in Latin America are served cold, as an appetizer, as a side dish or as a tapas dish in Spain. In several Central American countries it is prepared with cabbage, onions, carrots, lemon, vinegar, oregano, and salt. In Mexico, "curtido" consists of carrots, onions, and jalapeño peppers and used to accompany meals still common in taquerias and restaurants. In order to prepare a carrot "curtido" simply add carrots to vinegar and other ingredients that are common to the region such as chilli, tomato, and onions. Varies depending on the food, in the case of sour. Another example of a type of pickling which involves the pickling of meats or seafood is the "escabeche" or "ceviches" popular in Peru, Ecuador, and throughout Latin America and the Caribbean. These dishes include the pickling of pig's feet, pig's ears, and gizzards prepared as an "escabeche" with spices and seasonings to flavor it. The ceviches consists of shrimp, octopus, and various fishes seasoned and served cold.

Possible Health Hazards of Pickled Vegetables

The World Health Organization has listed pickled vegetables as a possible carcinogen, and the *British Journal of Cancer* released an online 2009 meta-analysis of research on pickles as increasing the risks of esophageal cancer. The report, citing limited data in a statistical meta analysis, indicates a potential two-fold increased risk of oesophageal cancer associated with Asian pickled vegetable consumption. Results from the research are described as having "high heterogeneity" and the study said that further well-designed prospective studies were warranted. However, their results stated "The majority of subgroup analyses showed a statistically significant association between consuming pickled vegetables and Oesophageal Squamous Cell Carcinoma".

The 2009 meta-analysis reported heavy infestation of pickled vegetables with fungi. Some common fungi can facilitate the formation of N-nitroso compounds, which are strong oesophageal carcinogens in several animal models. Roussin red methyl ester, a non-alkylating nitroso compound with tumour-promoting effect in vitro, was identified in pickles from Linxian in much higher concentrations than in samples from low-incidence areas. Fumonisin mycotoxins have been shown to cause liver and kidney tumours in rodents.

A 2017 study in *Chinese Journal of Cancer* has linked salted vegetables (common among Chinese cuisine) to a 4-fold increase in nasopharynx cancer, where fermentation was a critical step in creating nitrosamines, which some are confirmed carcinogens, as well as activation of Epstein–Barr virus by fermentation products.

Pickled cucumbers   pickled herring Pickled   mushrooms

Pickled olives   Pickled vegetables

Fermented homemade pickled cucumber, chili pepper, garlic, and apple in the hot climate of Indonesia

Sugaring

Sugaring is a method of food preservation that requires the food to be dehydrated and then to be packed with either crystallized sugar or with the liquids containing high amount of sugar such as honey or molasses.

The main purpose of this food preservation method is to treat the food in order to stop the growth of bacteria that may diminish the nutritional value and quality of the food. Some of the most popular sugared foods that are found in almost every cuisine across the globe include different kinds of fruits as well as a variety of vegetables such as ginger.

Although there is no specific evidence present to prove the origin of the sugaring as a food preservation method, the methods of preserving food adopted by the early American colonists used to include sugaring as one of the methods. Unlike today, early Americans were not so well-equipped with the techniques and means of transporting food from different climates and they could only rely upon the seasonal fruits and vegetables. With the lack of refrigeration, they were not able to store food for longer periods, therefore they developed various food preserving procedures that helped them in retaining the nutritional value and quality of food for extended period of time.

All the food preserving techniques had been introduced through trial and error, but the actual working of the methods was not known as bacteria had not been known till then. Though sugaring food preservation is believed to be done by the early Americans, the versatility of this food preservation made it popular in all cuisines.

The Process of Sugaring

Sugaring mainly reduces the action of the internal food enzymes by inhibiting the bacterial growth and this result in the preservation of food from spoiling.

The process of this type of food preservation is mainly done with any table or raw sugar that is generally in crystallized form. The sugar can also be liquefied in form of syrup, honey or molasses. These high sugar density liquids are of equal importance as sugar in the process of food preservation. Any fruit or vegetable that has to be preserved is washed thoroughly and desiccated by dehydration. The food is then cooked in liquid sugar product or raw sugar until crystallized and the resultant food is preserved in dried form.

Some fruits are even glazed in sugar syrup, but sold after being extracted from the liquid. The sugary coating on the fruits and the internal high sugar content increases the shelf life of the food.

Alcoholic preservation combined with sugar is also extremely popular to preserve some luxury items such as brandy and other spirits containing fruits. This type of food preservation should not be confused with the fruit flavored spirits like cherry brandy or apple wine.

Sugaring of Various Foods

- Fruits – Apples, pears, plum, cherry, apricots and peaches are some of the popular fruits that are commonly preserved by sugaring method. These fruits are either dried before preservation or glazed in sugar syrup. Fruits are also combined with alcohol and preserved.

- Vegetables – Ginger and carrot are the most common vegetables that are often sugared to prepare relishes or sweet pickles. These candied vegetables are popularly served as condiments.

- Angelica – It is a herb that is widely used as a flavoring agent. However, sugar preserved or candied strips of angelica are extremely popular as cake decorations.

- Citrus peel – The peels of citrus fruits like lemon and amla (Indian gooseberry) are often candied to form relishes. However, 'murabba' (Indian candied dish) includes whole amla.

Advantages and Disadvantages of Sugaring Food Preservation Method

Sugaring has few advantages over other preservation methods, as this process does not require large number of ingredients and often the sugar extract or glaze is used to sweeten various other foods as well. It is also an easy preservation method with less time involvement. There is a risk in this method as sugar is believed to attract moisture very fast. When the atmospheric moisture is high in content, the yeast present in the environment starts its action and sugar starts fermenting into carbon-di-oxide and alcohol. Although fermented food is also a preserved food, the sugared foods should be prevented from fermenting, as it may lead to an unpleasant taste.

Canning

Canning is the method of preserving food from spoilage by storing it in containers that are hermetically sealed and then sterilized by heat. The process was invented after prolonged research by Nicolas Appert of France in 1809, in response to a call by his government for a means of preserving food for army and navy use. Appert's method consisted of tightly sealing food inside a bottle or jar, heating it to a certain temperature, and maintaining the heat for a certain period, after which the container was kept sealed until use. It was 50 years before Louis Pasteur was able to explain why the food so treated did not spoil: the heat killed the microorganisms in the food, and the sealing kept other microorganisms from entering the jar. In 1810 Peter Durand of England patented the use of tin-coated iron cans instead of bottles, and by 1820 he was supplying canned food to the Royal Navy in large quantities. European canning methods reached the United States soon thereafter, and that country eventually became the world leader in both automated canning processes and total can production. In the late 19th century, Samuel C. Prescott and William Underwood of the United States set canning on a scientific basis by describing specific time-temperature heating requirements for sterilizing canned foods.

Originally, cans consisted of a sheet of tin-plated iron that was rolled into a cylinder (known as the body), onto which the top and bottom were manually soldered. This form was replaced in the early

20th century by the modern sanitary, or open-top, can, whose constituent parts are joined by interlocking folds that are crimped, or pressed together. Polymer sealing compounds are applied to the end, or lid, seams, and the body seams can be sealed on the outside by soldering. The modern tin can is made of 98.5 percent sheet steel with a thin coating of tin (i.e., tinplate). It is manufactured on wholly automatic lines of machinery at rates of hundreds of cans per minute.

Most vegetables, fruits, meat and dairy products, and processed foods are stored in tin cans, but soft drinks and many other beverages are now commonly stored in aluminum cans, which are lighter and do not rust. Aluminum cans are made by impact extrusion; the body of the can is punched out in one piece from a single aluminum sheet by a stamping die. This seamless piece, which has a rounded bottom, is then capped with a second piece as its lid. The tabs used in pop-top cans are also made of aluminum. Bimetal cans are made of aluminum bodies and steel lids.

Canneries are usually located close to the growing areas of the product to be packed, since it is desirable to can foods as quickly as possible after harvesting. The canning process itself consists of several stages: cleaning and further preparing the raw food material; blanching it; filling the containers, usually under a vacuum; closing and sealing the containers; sterilizing the canned products; and labeling and warehousing the finished goods. Cleaning usually involves passing the raw food through tanks of water or under high-pressure water sprays, after which vegetable or other products are cut, peeled, cored, sliced, graded, soaked, pureed, and so on. Almost all vegetables and some fruits require blanching by immersion in hot water or steam; this process softens the vegetable tissues and makes them pliable enough to be packed tightly, while also serving to inactivate enzymes that can cause undesirable changes in the food before canning. Blanching also serves as an additional or final cleansing operation.

The filling of cans is done automatically by machines; cans are filled with solid contents and, in many cases, with an accompanying liquid (often brine or syrup) in order to replace as much of the air in the can as possible. The filled cans are then passed through a hot-water or steam bath in an exhaust box; this heating expands the food and drives out the remaining air; thus, after sealing, heat sterilizing, and cooling the can, the contraction of the contents produces a partial vacuum within the container. Certain products are vacuum-packed, whereby the cans are mechanically exhausted by specially designed vacuum-can sealing machines.

Immediately after the cans are exhausted, they are closed and sealed; a machine places the cover on the can, and the curl on the can cover and the flange on the can body are rolled into position and then flattened together. The thin layer of sealing compound originally present in the rim of the cover is dispersed between the layers of metal to ensure a hermetic seal. The sealed cans are then sterilized; i.e., they are heated at temperatures high enough and for a long enough time to destroy all microorganisms (bacteria, molds, yeasts) that might still be present in the food contents. The heating is done in high-pressure steam kettles, or cookers, usually using temperatures around 240°F (116°C). The cans are then cooled in cold water or air, after which they are labeled.

Canning preserves most of the nutrients in foods. Proteins, carbohydrates, and fats are unaffected, as are vitamins A, C, D, and B2. The retention of vitamin B1 depends on the amount of heat used during canning. Some vitamins and minerals may dissolve into the brine or syrup in a can during processing, but they retain their nutritive value if those liquids are consumed.

There are two methods of canning, i.e., water-bath canning and pressure canning.

- Water-bath canning: This method, sometimes referred to as *hot water canning,* uses a large kettle of boiling water. Filled jars are submerged in the water and heated to an internal temperature of 212 degrees for a specific period of time. Use this method for processing high-acid foods, such as fruit, items made from fruit, pickles, pickled food, and tomatoes.

WATER-BATH CANNING KETTLE

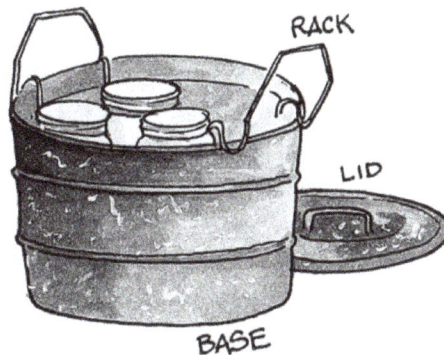

- Pressure canning: Pressure canning uses a large kettle that produces steam in a locked compartment. The filled jars in the kettle reach an internal temperature of 240 degrees under a specific pressure (stated in pounds) that's measured with a dial gauge or weighted gauge on the pressure-canner cover. Use a pressure canner for processing vegetables and other low-acid foods, such as meat, poultry, and fish.

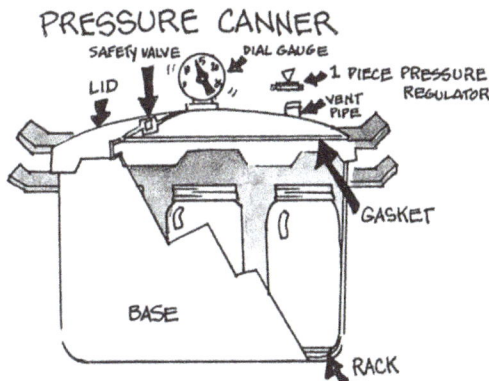

PRESSURE CANNER

Jellying

Food may be preserved by cooking in a material that solidifies to form a gel. Such materials include gelatin, agar, maize flour, and arrowroot flour. Some foods naturally form a protein gel when cooked, such as eels and elvers, and sipunculid worms, which are a delicacy in Xiamen, in the Fujian province of the People's Republic of China. Jellied eels are a delicacy in the East End of London, where they are eaten with mashed potatoes. Potted meats in aspic (a gel made from gelatin and clarified meat broth) were a common way of serving meat off-cuts in the UK until the 1950s. Many jugged meats are also jellied.

A traditional British way of preserving meat (particularly shrimp) is by setting it in a pot and sealing it with a layer of fat. Also common is potted chicken liver; jellying is one of the steps in producing traditional pates.

Burial

Burial of food can preserve it due to a variety of factors: lack of light, lack of oxygen, cool temperatures, pH level, or desiccants in the soil. Burial may be combined with other methods such as salting or fermentation. Most foods can be preserved in soil that is very dry and salty (thus a desiccant) such as sand, or soil that is frozen. Many root vegetables are very resistant to spoilage and require no other preservation than storage in cool dark conditions, for example by burial in the ground, such as in a storage clamp. Century eggs are created by placing eggs in alkaline mud (or other alkaline substance), resulting in their "inorganic" fermentation through raised pH instead of spoiling. The fermentation preserves them and breaks down some of the complex, less flavorful proteins and fats into simpler, more flavorful ones. Cabbage was traditionally buried in the fall in northern farms in the U.S. for preservation. Some methods keep it crispy while other methods produce sauerkraut. A similar process is used in the traditional production of kimchi. Sometimes meat is buried under conditions that cause preservation. If buried on hot coals or ashes, the heat can kill pathogens, the dry ash can desiccate, and the earth can block oxygen and further contamination. If buried where the earth is very cold, the earth acts like a refrigerator.

Fermentation

Fermentation in food processing typically is the conversion of carbohydrates to alcohols and carbon dioxide or organic acids using yeasts, bacteria, or a combination thereof, under anaerobic conditions. Fermentation in simple terms is the chemical conversion of sugars into ethanol. The science of fermentation is also known as zymology or zymurgy.

Historically, when studying the fermentation of sugar to alcohol by yeast, Louis Pasteur concluded that the fermentation was catalyzed by a vital force, called "ferments," within the yeast cells. The "ferments" were thought to function only within living organisms. "Alcoholic fermentation is an act correlated with the life and organization of the yeast cells, not with the death or putrefaction of the cells," he wrote.

Fermentation usually implies that the action of microorganisms is desirable. This process is used to produce alcoholic beverages such as wine, beer, and cider. Fermentation is also employed in the leavening of bread (CO_2 produced by yeast activity); in preservation techniques to produce lactic acid in sour foods such as sauerkraut, dry sausages, kimchi, and yogurt; and in the pickling of foods with vinegar (acetic acid).

Beer: This is beer fermenting at a brewery.

Food fermentation has been said to serve five main purposes:

1. Enrichment of the diet through development of a diversity of flavors, aromas, and textures in food substrates.

2. Preservation of substantial amounts of food through lactic acid, alcohol, acetic acid, and alkaline fermentations.

3. Biological enrichment of food substrates with protein, essential amino acids, essential fatty acids, and vitamins.

4. Elimination of antinutrients.

5. A decrease in cooking time and fuel requirement.

Process of Fermentation Preserves Foods

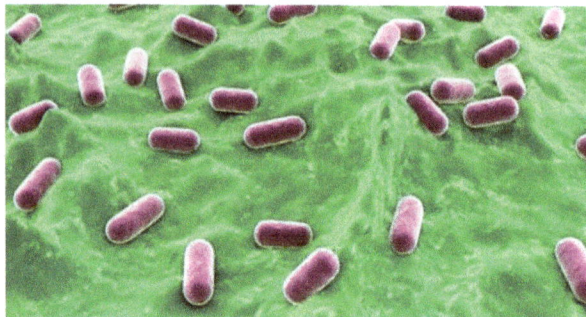

The desirable bacteria cause less deterioration of the food by inhibiting the growth of the spoiling types of bacteria. Some fermenting processes lower the pH of foods preventing harmful microorganisms to live with too acidic an environment. Controlled fermentation processes encourage the growth of good bacteria which starves, or fights off, the bad microbes. Depending on what is fermented, or the manner of fermentation, foods can remain consumable for years.

The fermentation process can be stopped by other means of preserving, such as, canning (heating), drying, or freezing. Heat (pasteurization, 145°F), and low temperatures (freezing, 32°F or below) stops the fermenting process by slowing, or killing, the preferred microorganisms, and other bacteria. A few undesirable bacteria are not killed by either means, and continue to grow. When the beneficial bacteria are gone, the unfavorable bacteria take over, growing exponentially! This causes rotting, disease, illness, and inedible foods. When the good guys are present and happy, the food remains edible.

Additional Benefits of Fermenting

Fermenting enhances the flavors of some foods, as with the extended fermentation of black teas, aged cheese, wine, and beer, which creates their distinctive flavors. Cocoa beans have to be fermented (composted) for a few days to remove the pods and to enhance the flavor of chocolate.

Fermenting makes foods more edible by changing chemical compounds, or predigesting, the foods for us. There are extreme examples of poisonous plants like cassava that are converted to edible products by fermenting. Some coffee beans are hulled by a wet fermenting process, as opposed to a dry process.

Fermentation increases nutritional values with the biochemical exchange it produces, and allows us to live healthier lives. Here are a few examples:

- The sprouting of grains, seeds, and nuts, multiplies the amino acid, vitamin, and mineral content and antioxidant qualities of the starting product.

- Fermented beans are easier for our bodies to digest, like the proteins found in soy beans that are nearly indigestible until fermented.

- Fermented dairy products, like, cheese, yogurt, and kifir, can be consumed by those not able to digest the raw milk, and aid the digestion and well-being for those with lactose intolerance and autism.

- Porridge made from grains allowed to ferment increases the nutritional values so much that it reduces the risk of disease in children.

- The news is full of reports about the health benefits of probotic supplements (beneficial bacterial cultures for microbial balance in the body) fighting cancer and other diseases.

- Vinegar is used to leach out certain flavors and compounds from plant materials to make healthy and tasty additions to our meals.

Pasteurization

Pasteurization is the process of heat processing a liquid or a food to kill pathogenic bacteria to make the food safe to eat. The use of pasteurization to kill pathogenic bacteria has helped reduce the transmission of diseases, such as typhoid fever, tuberculosis, scarlet fever, polio, and dysentery.

It is important to note that foods can become contaminated even after they have been pasteurized. For example, all pasteurized foods must be refrigerated. If the pasteurized food is temperature-abused (e.g., if milk or eggs are not kept refrigerated), it could become contaminated. Therefore, it is important to *always* handle food properly by handling it with clean hands, preventing it from becoming contaminated, and keeping it at a safe temperature.

Process of Pasteurization

Foods are heat-processed to kill pathogenic bacteria. Foods can also be pasteurized using gamma irradiation. Such treatments do not make the foods radioactive. The pasteurization process is based on the use of one of following time and temperature relationships.

High-Temperature-Short-Time Treatment (HTST) -- this process uses higher heat for less time to kill pathogenic bacteria. For example, milk is pasteurized at 161°F (72°C) for 15 seconds.

Low-Temperature-Long-Time Treatment (LTLT) -- this process uses lower heat for a longer time to kill pathogenic bacteria. For example, milk is pasteurized at 145°F (63°C) for 30 minutes.

It is important to remember that the times and temperatures depend on: (1) the type of food and (2) the final result one wants to achieve, such as retaining a food's nutrients, color, texture, and flavor.

Processes used to Pasteurize Foods

Flash Pasteurization - Involves a high-temperature, short-time treatment in which pourable products, such as juices, are heated for 3 to 15 seconds to a temperature that destroys harmful micro-organisms. After heating, the product is cooled and packaged. Most drink boxes and pouches use this pasteurization method as it allows extended unrefrigerated storage while providing a safe product.

Steam Pasteurization - This technology uses heat to control or reduce harmful microorganisms in beef. This system passes freshly-slaughtered beef carcasses that are already inspected, washed, and trimmed, through a chamber that exposes the beef to pressurized steam for approximately 6 to 8 seconds. The steam raises the surface temperature of the carcasses to 190° to 200°F (88° to 93°C). The carcasses are then cooled with a cold-water spray. This process has proven to be successful in reducing pathogenic bacteria, such as *E. coli* O157:H7, *Salmonella*, and *Listeria*, without the use of any chemicals. Steam pasteurization is used on nearly 50% of U.S. beef.

Irradiation Pasteurization - Foods, such as poultry, red meat, spices, and fruits and vegetables, are subjected to small amounts of gamma rays. This process effectively controls vegetative bacteria and parasitic foodborne pathogens and increases the storage time of foods.

The Effect of Pasteurization on Nutrients and Flavor

Pasteurization can affect the nutrient composition and flavor of foods. In the case of milk, for example, the high-temperature, short-time treatments (HTST) cause less damage to the nutrient composition and sensory characteristics of foods than do the low-temperature, long-time treatments (LTLT).

Examples of Foods that are Commonly Pasteurized

Whole Eggs Removed from Shells and Sold as a Liquid - Large quantities of eggs are sold to restaurants and institutions out of the shell. The yolk and whole-egg products are pasteurized in their raw form. The egg white is pasteurized in its raw form if it is sold as a liquid or frozen product.

Dried Eggs - If eggs are sold dried, the egg white with the glucose removed is normally heat-treated in the container by holding it for 7 days in a hot room at a minimum temperature of 130°F (54°C).

Whole Eggs Pasteurized in the Shell - Traditionally, eggs sold to customers in the shell have not been pasteurized. However, new time/temperature pasteurization methods are making this possible. Egg whites coagulate at 140°F (60°C). Therefore, heating an egg above 140°F would cook the egg, so processors pasteurize the egg in the shell at a low temperature, 130°F (54°C), for a long time, 45 minutes. This new process is being used by some manufacturers, but it is not yet widely available. Pasteurizing eggs reduces the risk of contamination from pathogenic bacteria, such as *Salmonella*, which can cause severe illness and even death. Pasteurized eggs in the shell may be used in recipes calling for raw eggs, such as Caesar salad, hollandaise or be arnaise sauces, mayonnaise, egg nog, ice cream, and egg-fortified beverages that are not thoroughly cooked.

Milk - Pasteurization improves the quality of milk and milk products and gives them a longer shelf life by destroying undesirable enzymes and spoilage bacteria. For example, the liquid is heated to 145°F (63°C) for at least 30 minutes or at least 161°F (72°C) for 15 seconds.

Today, many foods, such as eggs, milk, juices, spices and ice cream,are pasteurized. Sometimes higher temperatures are applied for a shorter period of time. The temperatures and times are determined by what is necessary to destroy pathogenic bacteria and other more heat-resistant disease-causing microorganisms that may be found in milk. The liquid is then quickly cooled to 40°F (4°C). Other liquids, such as juices, are heat-processed in a similar manner. Temperatures and times vary, depending on the product and the target organism.

Other Types of Milk Pasteurization

Ultrapasteurization - This involves the heating of milk and cream to at least 280° F (138° C) for at least 2 seconds, but because of less stringent packaging, they must be refrigerated. The shelf life of milk is extended 60 to 90 days. After opening, spoilage times for ultrapasteurized products are similar to those of conventionally pasteurized products.

Ultra-High-Temperature (UHT) Pasteurization -This involves heating milk or cream to 280° to 302°F (138° to 150°C) for 1 or 2 seconds. The milk is then packaged in sterile, hermetically-sealed (airtight) containers and can be stored without refrigeration for up to 90 days. After opening, spoilage times for UHT products are similar to those of conventionally pasteurized products.

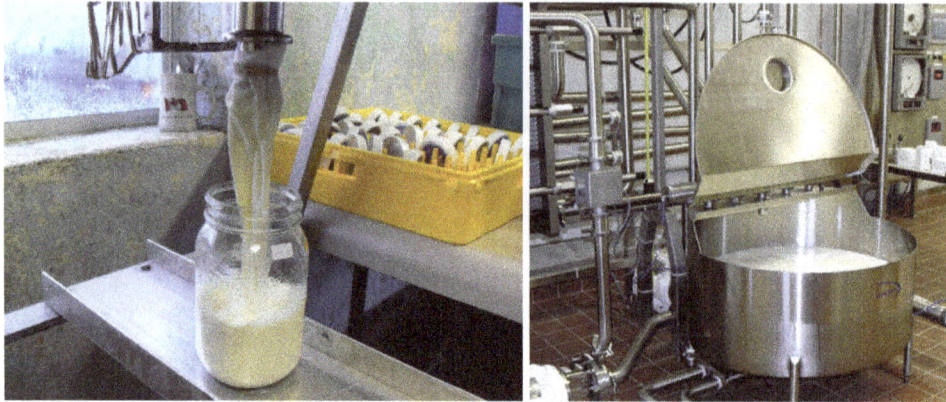

Microwave Volumetric Heating

Microwave heating refers to dielectric heating due to polarization effects at a selected frequency band in a nonconductor. Microwave heating in foods occurs due to coupling of electrical energy from an electromagnetic field in a microwave cavity with the food and its subsequent dissipation within food product. This results in a sharp increase in temperature within the product. Microwave energy is delivered at a molecular level through the molecular interaction with the electromagnetic field, in particular, through molecular friction resulting from dipole rotation of polar solvents and from the conductive migration of dissolved ions. Water in the food is the primary dipolar component responsible for the dielectric heating. In an alternating current electric field, the polarity of the field is varied at the rate of microwave frequency and molecules attempt to alignthemselves with the changing field. Heat is generated rapidly as a result of internal molecular friction.

The second major mechanism of heating with microwaves is through the polarization of ions as a result of the back and forth movement of the ionic molecules trying to align themselves with the oscillating electric field. (Oliveira and Franca 2002).

Advantages

i. Microwave penetrates inside the food materials and, therefore, cooking takes place throughout the whole volume of food internally, uniformly, and rapidly, which significantly reduces the processing time and energy.

ii. Since the heat transfer is fast, nutrients and vitamins contents, as well as flavor, sensory characteristics, and color of food are well preserved.

iii. Minimum fouling depositions, because of the elimination of the hot heat transfer surfaces, since the piping used is microwave transparent and remains relatively cooler than the product.

iv. High heating efficiency (80% or higher efficiency can be achieved).

v. Perfect geometry for clean-in-place (CIP) system.

vi. Suitable for heat-sensitive, high-viscous.

vii. Multiphase fluids.

Factors Affecting Microwave Heating

Some physical, thermal, and electrical properties determine the absorption of microwave energy and simultaneous heating behavior of food materials in microwave processing. These properties/factors are briefly discussed below.

Frequency

For food application, only two frequencies are allocated for microwave heating (915 and 2450 MHz) and therefore, these frequencies are of special interest. The corresponding wavelengths of these frequencies are 0.328 and 0.122 m, respectively. The wavelength has special significance as most interactions between the energy and materials take place in that region and generate instantaneous heat due to molecular friction. Food constituents except moisture, lipids, and ash are relatively inert to prescribed microwave frequencies. In addition, frequency (or wavelength) dictates equipment components such as magnetron, waveguide, and to some extent heating volume.

Dielectric Properties

The electrical properties of materials in the context of microwave and radiofrequency heating are known as dielectric properties, which provide a measure of how food materials interact with electromagnetic energy. Biological materials may be viewed as non ideal capacitors in that they have the ability to store and dissipate electrical energy from an electromagnetic field and the properties can be expressed in terms of a complex notation. The complex notation is characterized by dielectric permittivity with a real component, dielectric constant, and an imaginary component, dielectric loss.

Moisture Content

The moisture content significantly affects the dielectric properties of the food product and consequently penetration depth of the microwaves. Uneven heating rate is observed in high-moisture foods because of low microwave penetration depth. Low-moisture foods will have more uniform heating rate because of the deeper microwave penetration. The initial moisture content of the product and the rate of moisture evaporation play important roles during microwave heating. The heating behavior of water is phase dependent (liquid water versus solid ice phase) and also depends on the available free water content.

Mass

A direct relationship exists between the mass and the amount of absorbed microwave power, which should be applied to achieve the desired heating. For a smaller mass, batch oven is suitable, while a larger through put would often be better in large capacity equipment with conveyor. Such equipment has the added advantage of providing greater heating uniformity by moving the product through the microwave field. Each microwave oven has a critical (minimum) sample mass for its efficient operation. It is usually around250 ml water load in a 1 kW oven. Below this level, significant amount of microwave power is not absorbed into the product, and at very low loads they may damage the magnetron.

Temperature

Microwave heating is significantly affected by the level of sample temperature. Dielectric properties may vary with temperature, depending upon the material. Both temperature and moisture content can change during heating and therefore, those may have a combined effect on the dielectric constant, dielectric loss factor, loss tangent, and subsequently on the heating behavior. The initial temperature of the food product being heated by microwaves should either be controlled or known, so that the microwave power can be adjusted to obtain uniform final temperatures. If the microwave oven is preset to increase the product temperature from 20°C to 80°C, it will practically reach target temperature of 95°C with an initial product temperature of 35°C. To compensate the effect of higher initial temperature, the power of MW oven should be reduced or a higher sample mass should be used or the product should be heated for a shorter duration.

Geometry and Location of Foods

The shape of the food product does play an important role in the distribution of heat within the product heated in a microwave oven. It affects the depth of microwave penetration, and the heating rate and uniformity. Irregular-shaped products are subjected to non uniform heating due to the difference in product thickness. The closer the size (thickness) is to the wavelength, the higher will be the center temperature. Smaller particulates require less heat than larger ones. In addition, the more regular the shape, the more uniform will be the heat distribution within the product. A food of a spherical or cylindrical shape heats more evenly than a square. A higher surface-to-volume ratio enhances the heating rate. Therefore, the heating rate for a sphere will be different from that of a cylinder with the same volume. Placement has the most significant effect.

Thermal Properties

The heating characteristics of foods are dependent to a greater or lesser extent on some thermal properties such as thermal conductivity, density, and heat capacity. Thermal conductivity of food plays a significant role in microwave heating. Materials with higher thermal conductivity dissipate heat faster than the ones with lower conductivity during microwave heating. Food with high thermal conductivity will take lesser time to attain uniform temperature during holding.

Application in Food Industry

The major industrial applications of microwave heating are tempering of frozen meat and poultry products; precooking of bacon for foodservice; sausage cooking; drying of various foods; baking of bread, biscuit, and confectionery; thawing of frozen products; blanching of vegetables; heating and sterilizing of fast food, cooked meals, and cereals; and pasteurization and sterilization of various foods.

Application in Dairy Industry

Milk is traditionally pasteurized in a heat exchanger before distribution. The application of microwave heating to pasteurize milk has been well studied and has been a commercial practice for quite a long time. The success of microwave heating of milk is based on established conditions that provide the desired degree of safety with minimum product quality degradation. Since the first reported study on the use of a microwave system for pasteurization of milk (Hamid et al 1969), several studies on microwave heating of milk have been carried out. The majority of these microwave-based studies have been used to investigate the possibility of shelf-life enhancement of pasteurized milk, application of microwave energy to inactivate milk pathogens, assess the influence on the milk nutrients or the non-uniform temperature distribution during the microwave treatment.

Irradiation

Food irradiation is a processing and preservation technique with similar results to freezing or pasteurisation. During this procedure, the food is exposed to doses of ionising energy, or radiation. At low doses, irradiation extends a product's shelf life. At higher doses, this process kills insects, moulds, bacteria and other potentially harmful micro-organisms.

Considerable scientific research over the past five decades indicates that food irradiation is a safe and effective form of processing. Food irradiation has been approved in 40 countries including Australia, the United States, Japan, China, France and Holland.

To date, in Australia and New Zealand, only herbs and spices, herbal infusions and some tropical fruits are approved for irradiation by Food Standards Australia New Zealand (FSANZ), in accordance with the FSANZ Food Standards Code. For each of these, FSANZ has established that there are no safety concerns and no significant nutritional changes to the food as a result of food irradiation. Irradiated foods will be clearly labelled so that consumers can make an informed choice.

Irradiated Foods and Radioactivity

There is a common misconception that irradiated food is radioactive. The radiation used to process foods is very different from the radioactive fallout that occurs after, for example, a nuclear accident.

In food processing, the radioactive sources permitted do not generate gamma, electrons or x-rays of sufficient high energy to make food radioactive. No radioactive energy remains in the food after treatment.

The World Health Organization (WHO), the American Dietetic Association and the Scientific Committee of the European Union are three internationally recognised bodies that support food irradiation.

Food Irradiation Procedure

The food is exposed to ionising radiation, either from gamma rays or a high-energy electron beam or powerful x-rays. Gamma rays and x-rays are a form of radiation that shares some characteristics with microwaves, but with much higher energy and penetration.

The rays pass through the food just like microwaves in a microwave oven, but the food does not heat up to any significant extent. Exposure to gamma rays does not make food radioactive. Electron beams and x-rays are produced using electricity, which can be switched on or off, and they do not require radioactive material.

In both cases, organisms that are responsible for spoiling foods— such as insects, moulds and bacteria, including some important food poisoning bacteria— can be killed. Food irradiation cannot kill viruses.

Effects of Irradiation on Food

Some foods, such as dairy foods and eggs, cannot be irradiated because it causes changes in flavour or texture. Fruits, vegetables, grain foods, spices and meats (such as chicken) can be irradiated.

Irradiation causes minimal changes to the chemical composition of the food, however, it can alter the nutrient content of some foods because it reduces the level of some of the B-group vitamins. This loss is similar to those that occur when food is cooked or preserved in more traditional and accepted ways, such as canning or blanching.

Labelling of Irradiated Foods

If a food has been irradiated or contains irradiated ingredients or components, it must be labelled with a statement that the food, ingredients or components have been treated with ionising radiation.

If a food product does not have a label (such as whole fruits sold loose), this statement must be displayed in close proximity to the food. In addition to mandatory labelling, irradiated foods may also be marked with a symbol called a 'radura', which is the international symbol for irradiation.

Proper Food Handling is Still Needed

Food irradiation can only be used if it fulfils a technological need or is necessary for a food safety or food hygiene purpose. It does not replace the need for correct food handling practices in industry and in the home. For instance, a few bacteria may survive the irradiation of meat. If the meat is left unrefrigerated, these bacteria could still multiply and cause food poisoning.

Uses

Irradiation is used to reduce or eliminate the risk of food-borne illnesses, prevent or slow down spoilage, arrest maturation or sprouting and as a treatment against pests. Depending on the dose, some or all of the pathogenic organisms, microorganisms, bacteria, and viruses present are destroyed, slowed down, or rendered incapable of reproduction. Irradiation cannot revert spoiled or over ripened food to a fresh state. If this food was processed by irradiation, further spoilage would cease and ripening would slow down, yet the irradiation would not destroy the toxins or repair the texture, color, or taste of the food. When targeting bacteria, most foods are irradiated to significantly reduce the number of active microbes, not to sterilize all microbes in the product. In this respect it is similar to pasteurization.

Irradiation is used to create safe foods for people at high risk of infection, or for conditions where food must be stored for long periods of time and proper storage conditions are not available. Foods that can tolerate irradiation at sufficient doses are treated to ensure that the product is completely sterilized. This is most commonly done with rations for astronauts, and special diets for hospital patients.

Irradiation is used to create shelf-stable products. Since irradiation reduces the populations of spoilage microorganisms, and because pre-packed food can be irradiated, the packaging prevents recontamination of the final product.

Irradiation is used to reduce post-harvest losses. It reduces populations of spoilage micro-organisms in the food and can slow down the speed at which enzymes change the food, and therefore slows spoilage and ripening, and inhibits sprouting (e.g., of potato, onion, and garlic).

Food is also irradiated to prevent the spread of invasive pest species through trade in fresh vegetables and fruits, either within countries, or trade across international boundaries. Pests such as insects could be transported to new habitats through trade in fresh produce which could significantly affect agricultural production and the environment were they to establish themselves. This "phytosanitary irradiation" aims to render any hitch-hiking pest incapable of breeding. The pests are sterilized when the food is treated by low doses of irradiation. In general, the higher doses required to destroy pests such as insects, mealybugs, mites, moths, and butterflies either affect the look or taste, or cannot be tolerated by fresh produce. Low dosage treatments (less than 1000 gray) enables trade across quarantine boundaries and may also help reduce spoilage.

Impact

Irradiation reduces the risk of infection and spoilage, does not make food radioactive, and the food is shown to be safe, but it does cause chemical reactions that alter the food and therefore alters the chemical makeup, nutritional content, and the sensory qualities of the food. Some of the potential secondary impacts of irradiation are hypothetical, while others are demonstrated. These effects include cumulative impacts to pathogens, people, and the environment due to the reduction of food quality, the transportation and storage of radioactive goods, and destruction of pathogens, changes in the way we relate to food and how irradiation changes the food production and shipping industries.

Immediate Effects

The radiation source supplies energetic particles or waves. As these waves/particles pass through a target material they collide with other particles. Around the sites of these collisions chemical bonds are broken, creating short lived radicals (e.g. the hydroxyl radical, the hydrogen atom and solvated electrons). These radicals cause further chemical changes by bonding with and or stripping particles from nearby molecules. When collisions damage DNA or RNA, effective reproduction becomes unlikely, also when collisions occur in cells, cell division is often suppressed.

Irradiation (within the accepted energy limits, as 10 MeV for electrons, 5 MeV for X-rays [US 7.5 MeV] and gamma rays from Cobalt-60) can not make food radioactive, but it does produce radiolytic products, and free radicals in the food. A few of these products are unique, but not considered dangerous.

Irradiation can also alter the nutritional content and flavor of foods, much like cooking. The scale of these chemical changes is not unique. Cooking, smoking, salting, and other less novel techniques, cause the food to be altered so drastically that its original nature is almost unrecognizable, and must be called by a different name. Storage of food also causes dramatic chemical changes, ones that eventually lead to deterioration and spoilage.

Misconceptions

A major concern is that irradiation might cause chemical changes that are harmful to the consumer. Several national expert groups and two international expert groups evaluated the available data and concluded that any food at any dose is wholesome and safe to consume as long as it remains palatable and maintains its technical properties (e.g. feel, texture, or color).

Irradiated food does not become radioactive, just as an object exposed to light does not start producing light. Radioactivity is the ability of a substance to emit high energy particles. When particles hit the target materials they may free other highly energetic particles. This ends shortly after the end of the exposure, much like objects stop reflecting light when the source is turned off and warm objects emit heat until they cool down but do not continue to produce their own heat. To modify a material so that it keeps emitting radiation (induce radiation) the atomic cores (nucleus) of the atoms in the target material must be modified.

It is impossible for food irradiators to induce radiation in a product. Irradiators emit electrons or photons and the radiation is intrinsically radiated at precisely known strengths (wavelengths for photons, and speeds for electrons). These radiated particles at these strengths can never be strong enough to modify the nucleus of the targeted atom in the food, regardless of how many particles hit the target material, and radioactivity can not be induced without modifying the nucleus.

Chemical Changes

Compounds known as free radicals form when food is irradiated. Most of these are oxidizers (i.e., accept electrons) and some react very strongly. According to the free-radical theory of aging excessive amounts of these free radicals can lead to cell injury and cell death, which may contribute to many diseases. However, this generally relates to the free radicals generated in the body, not the free radicals consumed by the individual, as much of these are destroyed in the digestive process.

When fatty acids are irradiated, a family of compounds called 2-alkylcyclobutanones (2-ACBs) are produced. These are thought to be unique radiolytic products. Most of the substances found in irradiated food are also found in food that has been subjected to other food processing treatments, and are therefore not unique. One family of chemicals (2ACB's) are uniquely formed by irradiation (unique radiolytic products), and this product is nontoxic. When irradiating food, all other chemicals occur in a lower or comparable frequency to other food processing techniques. Furthermore, the quantities in which they occur in irradiated food are lower or similar to the quantities formed in heat treatments.

The radiation doses to cause toxic changes are much higher than the doses used to during irradiation, and taking into account the presence of 2-ACBs along with what is known of free radicals, these results lead to the conclusion that there is no significant risk from radiolytic products.

Food Quality

Ionizing radiation can change food quality but in general very high levels of radiation treatment (many thousands of gray) are necessary to adversely change nutritional content, as well as the sensory qualities (taste, appearance, and texture). Irradiation to the doses used commercially to treat food have very little negative impact on the sensory qualities and nutrient content in foods. When irradiation is used to maintain food quality for a longer period of time (improve the shelf stability of some sensory qualities and nutrients) the improvement means that more consumers have access to the original taste, texture, appearance, and nutrients. The changes in quality and nutrition depend on the degree of treatment and may vary greatly from food to food.

There has been low level gamma irradiation that has been attempted on arugula, spinach, cauliflower, ash gourd, bamboo shoots, coriander, parsley, and watercress. There has been limited information, however, regarding the physical, chemical and/or bioactive properties and the shelf life on these minimally processed vegetables.

There is some degradation of vitamins caused by irradiation, but is similar to or even less than the loss caused by other processes that achieve the same result. Other processes like chilling, freezing, drying, and heating also result in some vitamin loss.

The changes in the flavor of fatty foods like meats, nuts and oils are sometimes noticeable, while the changes in lean products like fruits and vegetables are less so. Some studies by the irradiation industry show that for some properly treated fruits and vegetables irradiation is seen by consumers to improve the sensory qualities of the product compared to untreated fruits and vegetables.

Quality Impact on Minimally Processed Vegetables

Watercress (*Nasturtium Officinale*) is a rapidly growing aquatic or semi aquatic perennial plant. Because chemical agents do not provide efficient microbial reductions, watercress has been tested with gamma irradiation treatment in order to improve both safety and the shelf life of the product. It is traditionally used on horticultural products to prevent sprouting and post-packaging contamination, delay post-harvest ripening, maturation and senescence.

In a Food Chemistry food journal, scientists studied the suitability of gamma irradiation of 1, 2, and 5 kGy for preserving quality parameters of the fresh cut watercress at around 4 degrees Celsius for 7 days. They determined that a 2 kGy dose of irradiation was the dose that contained most similar qualities to non-stored control samples, which is one of the goals of irradiation. 2 kGy preserved high levels of reducing sugars and favoured PUFA; while samples of the 5 kGy dose revealed high contents of sucrose and MUFA. Both cases the watercress samples obtained healthier fatty acids profiles. However, a 5kGy dose better preserved the antioxidant activity and total flavonoids.

Long-term Impacts

If the majority of food was irradiated at high-enough levels to significantly decrease its nutritional content, there would be an increased risk of developing nutritionally-based illnesses if additional steps, such as changes in eating habits, were not taken to mitigate this. Furthermore, for at least three studies on cats, the consumption of irradiated food was associated with a loss of tissue in the myelin sheath, leading to reversible paralysis. Researchers suspect that reduced levels of vitamin A and high levels of free radicals may be the cause. This effect is thought to be specific to cats and has not been reproduced in any other animal. To produce these effects, the cats were fed solely on food that was irradiated at a dose at least five times higher than the maximum allowable dose.

It may seem reasonable to assume that irradiating food might lead to radiation-tolerant strains, similar to the way that strains of bacteria have developed resistance to antibiotics. Bacteria develop a resistance to antibiotics after an individual uses antibiotics repeatedly. Much like pasteurization plants, products that pass through irradiation plants are processed once, and are not processed and reprocessed. Cycles of heat treatment have been shown to produce heat-tolerant bacteria, yet no problems have appeared so far in pasteurization plants. Furthermore, when the irradiation dose is chosen to target a specific species of microbe, it is calibrated to doses several times the value required to target the species. This ensures that the process randomly destroys all members of a target species. Therefore, the more irradiation-tolerant members of the target species are not given any evolutionary advantage. Without evolutionary advantage, selection does not occur. As to the irradiation process directly producing mutations that lead to more virulent, radiation-resistant strains, the European Commission's Scientific Committee on Food found that there is no evidence; on the contrary, irradiation has been found to cause loss of virulence and infectivity, as mutants are usually less competitive and less adapted.

Misconceptions

Some who advocate against food irradiation argue the safety of irradiated food is not scientifically proven because there are a lack of long-term studies in spite of the fact that hundreds of animal feeding studies of irradiated food, including multigenerational studies, have been performed since 1950. Endpoints investigated have included sub chronic and chronic changes in metabolism, histopathology, function of most systems, reproductive effects, growth, teratogenicity, and mutagenicity. A large number of studies have been performed; meta-studies have supported the safety of irradiated food.

The below experiments are cited by food irradiation opponents, but either could not be verified in later experiments, could not be clearly attributed to the radiation effect, or could be attributed to an inappropriate design of the experiment.

- India's National Institute of Nutrition (NIN) found an elevated rate of cells with more than one set of genes (polyploidy) in humans and animals when fed wheat that was irradiated recently (within 12 weeks). Upon analysis, scientists determined that the techniques used by the NIN allowed for too much human error and statistical variation; therefore, the results were unreliable. After multiple studies by independent agencies and scientists, no correlation between polyploidy and irradiation of food could be found.

Indirect Effects of Irradiation

The indirect effects of irradiation are the concerns and benefits of irradiation that are related to how making food irradiation a common process will change the world, with emphasis on the system of food production.

If irradiation was to become common in the food handling process there would be a reduction of the prevalence of foodborne illness and potentially the eradication of specific pathogens. However, multiple studies suggest that an increased rate of pathogen growth may occur when irradiated food is cross-contaminated with a pathogen, as the competing spoilage organisms are no longer present. This being said, cross contamination itself becomes less prevalent with an increase in usage of irradiated foods.

The ability to remove bacterial contamination through post-processing by irradiation may reduce the fear of mishandling food which could cultivate a cavalier attitude toward hygiene and result in contaminants other than bacteria. However, concerns that the pasteurization of milk would lead to increased contamination of milk were prevalent when mandatory pasteurization was introduced, but these fears never materialized after adoption of this law. Therefore, it is unlikely for irradiation to cause an increase of illness due to non bacteria-based contamination.

Treatment

Up to the point where the food is processed by irradiation, the food is processed in the same way as all other food. To treat the food, they are exposed to a radioactive source, for a set period of time to achieve a desired dose. Radiation may be emitted by a radioactive substance, or by X-ray and electron beam accelerators. Special precautions are taken to ensure the food stuffs never come in contact with the radioactive substances and that the personnel and the environment are protected from exposure radiation.Irradiation treatments are typically classified by dose (high, medium, and low), but are sometimes classified by the effects of the treatment (radappertization, radicidation and radurization). Food irradiation is sometimes referred to as "cold pasteurization" or "electronic pasteurization" because ionizing the food does not heat the food to high temperatures during the process, and the effect is similar to heat pasteurization. The term "cold pasteurization" is controversial because the term may be used to disguise the fact the food has been irradiated and pasteurization and irradiation are fundamentally different processes.

Treatment costs vary as a function of dose and facility usage. A pallet or tote is typically exposed for several minutes to hours depending on dose. Low-dose applications such as disinfestation of fruit range between US$0.01/lbs and US$0.08/lbs while higher-dose applications can cost as much as US$0.20/lbs.

Packaging

Food processors and manufacturers today struggle with using affordable, efficient packaging materials for irradiation based processing. The implementation of irradiation on prepackaged foods has been found to impact foods by inducing specific chemical alterations to the food packaging material that migrates into the food. Cross-linking in various plastics can lead to physical and chemical modifications that can increase the overall molecular weight. On the other hand, chain scission is fragmentation of polymer chains that leads to a molecular weight reduction.

Dosimetry

The radiation absorbed dose is the amount energy absorbed per unit weight of the target material. Dose is used because, when the same substance is given the same dose, similar changes are observed in the target material. The SI unit for dose is grays (Gy or J/kg). Dosimeters are used to measure dose, and are small components that, when exposed to ionizing radiation, change measurable physical attributes to a degree that can be correlated to the dose received. Measuring dose (dosimetry) involves exposing one or more dosimeters along with the target material.

For purposes of legislation doses are divided into low (up to 1 kGy), medium (1 kGy to 10 kGy), and high-dose applications (above 10 kGy). High-dose applications are above those currently permitted in the US for commercial food items by the FDA and other regulators around the world. Though these doses are approved for non commercial applications, such as sterilizing frozen meat for NASA astronauts (doses of 44 kGy) and food for hospital patients.

Applications of food irradiation		
	Application	Dose (kGy)
Low dose (up to 1 kGy)	Inhibit sprouting (potatoes, onions, yams, garlic)	0.06 - 0.2
	Delay in ripening (strawberries, potatoes)	0.5 - 1.0
	Prevent insect infestation (grains, cereals, coffee beans, spices, dried nuts, dried fruits, dried fish, mangoes, papayas)	0.15 - 1.0
	Parasite control and inactivation (tape worm, trichina)	0.3 - 1.0
Medium dose (1 kGy to 10 kGy)	Extend shelf-life (raw and fresh fish, seafood, fresh produce, refrigerated and frozen meat products)	1.0 - 7.0
	Reduce risk of pathogenic and spoilage microbes (meat, seafood, spices, and poultry)	1.0 - 7.0
	Increased juice yield, reduction in cooking time of dried vegetables	3.0 - 7.0
High dose (above 10 kGy)	Enzymes (dehydrated)	10.0
	Sterilization of spices, dry vegetable seasonings	30.0 max
	Sterilization of packaging material	10.0 - 25.0
	Sterilization of foods (NASA and hospitals)	44.0

Processes

Gamma Irradiation

Gamma irradiation is produced from the radioisotopes cobalt-60 and caesium-137, which are derived by neutron bombardment of cobalt-59 and as a nuclear source by-product, respectively. Cobalt-60 is the most common source of gamma rays for food irradiation in commercial scale facilities as it is water insoluble and hence has little risk of environmental contamination by leakage into the water systems. As for transportation of the radiation source, cobalt-60 is transported in special trucks that prevent release of radiation and meet standards mentioned in the Regulations for Safe Transport of Radioactive Materials of the International Atomic Energy Act. The special trucks must meet high safety standards and pass extensive tests to be approved to ship radiation sources. Conversely, caesium-137, is water soluble and poses a risk of environmental contamination. Insufficient quantities are available for large scale commercial use. An incident where water-soluble caesium-137 leaked into the source storage pool requiring NRC intervention has led to near elimination of this radioisotope.

Cobalt 60 stored in Gamma Irradiation machine

Gamma irradiation is widely used due to its high penetration depth and dose uniformity, allowing for large-scale applications with high through puts. Additionally, gamma irradiation is significantly less expensive than using an X-ray source In most designs, the radioisotope, contained in stainless steel pencils, is stored in a water-filled storage pool which absorbs the radiation energy when not in use. For treatment, the source is lifted out of the storage tank, and product contained in totes is passed around the pencils to achieve required processing.

Electron Beam

Treatment of electron beams is created as a result of high energy electrons in an accelerator that generates electrons accelerated to 99% the speed of light. This system uses electrical energy and can be powered on and off. The high power correlates with a higher throughput and lower unit cost, but electron beams have low dose uniformity and a penetration depth of centimeters. Therefore, electron beam treatment works for products that have low thickness.

Irradiated Guava: Spring Valley Fruits, Mexico

X-ray

X-rays are produced by bombardment of dense target material with high energy accelerated electrons(this process is known as bremsstrahlung-conversion), giving rise to a continuous energy spectrum. Heavy metals, such as tantalum and tungsten, are used because of their high atomic numbers and high melting temperatures. Tantalum is usually preferred versus tungsten for industrial, large-area, high-power targets because it is more workable than tungsten and has a higher threshold energy for induced reactions. Like electron beams, x-rays do not require the use of radioactive materials and can be turned off when not in use. X-rays have high penetration depths and high dose uniformity but they are a very expensive source of irradiation as only 8% of the incident energy is converted into X-rays.

Cost

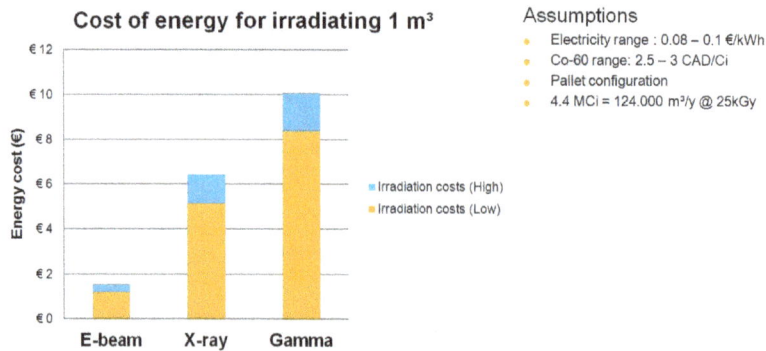

Efficiency illustration of the different radiation technologies (electron beam, X-ray, gamma rays)

The cost of food irradiation is influenced by dose requirements, the food's tolerance of radiation, handling conditions, i.e., packaging and stacking requirements, construction costs, financing arrangements, and other variables particular to the situation. Irradiation is a capital-intensive technology requiring a substantial initial investment, ranging from $1 million to $5 million. In the case of large research or contract irradiation facilities, major capital costs include a radiation source, hardware (irradiator, totes and conveyors, control systems, and other auxiliary equipment), land (1 to 1.5 acres), radiation shield, and warehouse. Operating costs include salaries (for fixed and variable labor), utilities, maintenance, taxes/insurance, cobalt-60 replenishment, general utilities, and miscellaneous operating costs. Perishable food items, like fruits, vegetables and meats would still require to be handled in the cold chain, so all other supply chain costs remain the same.

Public Perception

Negative connotations associated with the word "radiation" are thought to be responsible for low consumer acceptance. Several national expert groups and two international expert groups evaluated the available data and concluded that any food at any dose is wholesome and safe to consume.

Irradiation has been approved by many countries. For example, in the U.S. the FDA has approved food irradiation for over fifty years. However, in the past decade the major growth area is for fruits and vegetables that are irradiated to prevent the spread of pests. In the early 2000s in the US, irradiated meat was common at some grocery stores, but because of lack of consumer demand, it is no longer common. Because consumer demand for irradiated food is low, reducing the spoilage between manufacturer and consumer purchase and reducing the risk of food borne illness is currently not sufficient incentive for most manufacturers to supplement their process with irradiation. Nevertheless, food irradiation does take place commercially and volumes are in general increasing at a slow rate, even in the European Union where all member countries allow the irradiation of dried herbs spices and vegetable seasonings but only a few allow other foods to be sold as irradiated.

Although there are some consumers who choose not to purchase irradiated food, a sufficient market has existed for retailers to have continuously stocked irradiated products for years. When labeled irradiated food is offered for retail sale, these consumers buy it and re-purchase it, indicating that it is possible to successfully market irradiated foods, therefore retailers not stocking irradiated foods might be a major bottleneck to the wider adoption of irradiated foods. It is however, widely believed that consumer perception of foods treated with irradiation is more negative than those processed by other means and some industry studies indicate the number of consumers concerned about the safety of irradiated food decreased between 1985 and 1995 to levels comparable to those of people concerned about food additives and preservatives. Even though is it is untrue "People think the product is radioactive," said Harlan Clemmons, president of Sadex, a food irradiation company based in Sioux City, Iowa. Because of these concerns and the increased cost of irradiated foods, there is not a widespread public demand for the irradiation of foods for human consumption. Irradiated food does not become radioactive.

Standards and Regulations

The Codex Alimentarius represents the global standard for irradiation of food, in particular under the WTO-agreement. Regardless of treatment source, all processing facilities must adhere to safety standards set by the International Atomic Energy Agency (IAEA), Codex Code of Practice for the Radiation Processing of Food, Nuclear Regulatory Commission (NRC), and the International Organization for Standardization (ISO). More specifically, ISO 14470 and ISO 9001 provide in-depth information regarding safety in irradiation facilities.

All commercial irradiation facilities contain safety systems are designed to prevent exposure of personnel to radiation. The radiation source is constantly shielded by water, concrete, or metal. Irradiation facilities are designed with overlapping layers of protection, interlocks, and safeguards to prevent accidental radiation exposure. Additionally, "melt-downs" do not occur in facilities because the radiation source gives off radiation and decay heat; however, the heat is not sufficient to melt any material.

Labeling

The provisions of the Codex Alimentarius are that any "first generation" product must be labeled "irradiated" as any product derived directly from an irradiated raw material; for ingredients the provision is that even the last molecule of an irradiated ingredient must be listed with the ingredients even in cases where the unirradiated ingredient does not appear on the label. The RADURA-logo is optional; several countries use a graphical version that differs from the Codex-version. The suggested rules for labeling is published at CODEX-STAN – 1 (2005), and includes the usage of the Radura symbol for all products that contain irradiated foods. The Radura symbol is not a designator of quality. The amount of pathogens remaining is based upon dose and the original content and the dose applied can vary on a product by product basis.

The European Union follows the Codex's provision to label irradiated ingredients down to the last molecule of irradiated food. The European Community does not provide for the use of the Radura logo and relies exclusively on labeling by the appropriate phrases in the respective languages of the Member States. The European Union enforces its irradiation labeling laws by requiring its member countries to perform tests on a cross section of food items in the market-place and to report to the European Commission. The results are published annually in the OJ of the European Communities.

The Radura symbol, as required by U.S. Food and Drug Administration regulations
to show a food has been treated with ionizing radiation.

The US defines irradiated foods as foods in which the irradiation causes a material change in the food, or a material change in the consequences that may result from the use of the food. Therefore, food that is processed as an ingredient by a restaurant or food processor is exempt from the labeling requirement in the US. All irradiated foods must include a prominent Radura symbol followed in addition to the statement "treated with irradiation" or "treated by irradiation. Bulk foods must be individually labeled with the symbol and statement or, alternatively, the Radura and statement should be located next to the sale container.

Packaging

Under section 409 of the Federal Food, Drug, and Cosmetic Act, irradiation of prepackaged foods requires premarket approval for not only the irradiation source for a specific food but also for the food packaging material. Approved packaging materials include various plastic films, yet does not cover a variety of polymers and adhesive based materials that have been found to meet specific standards. The lack of packaging material approval limits manufacturers production and expansion of irradiated prepackaged foods.

Approved materials by FDA for Irradiation according to 21 CFR 179.45:

Material	Paper (kraft)	Paper (glassine)	Paper-board	Cello-phane (coated)	Polyolefin film	Polyestyrene film	Ny-lon-6	Vegetable Parchment	Nylon 11
Irradiation (kGy)	.05	10	10	10	10	10	10	60	60

Food Safety

In 2003, the Codex Alimentarius removed any upper dose limit for food irradiation as well as clearances for specific foods, declaring that all are safe to irradiate. Countries such as Pakistan and Brazil have adopted the Codex without any reservation or restriction. Other countries, including New Zealand, Australia, Thailand, India, and Mexico, have permitted the irradiation of fresh fruits for fruit fly quarantine purposes, amongst others.

Standards that describe calibration and operation for radiation dosimetry, as well as procedures to relate the measured dose to the effects achieved and to report and document such results, are maintained by the American Society for Testing and Materials (ASTM international) and are also available as ISO/ASTM standards.

All of the rules involved in processing food are applied to all foods before they are irradiated.

The U.S. Food and Drug Administration (FDA) is the agency responsible for regulation of radiation sources in the United States. Irradiation, as defined by the FDA is a "food additive" as opposed to a food process and therefore falls under the food additive regulations. Each food approved for irradiation has specific guidelines in terms of minimum and maximum dosage as deterred safe by the FDA. Packaging materials containing the food processed by irradiation must also undergo approval. The United States Department of Agriculture (USDA) amends these rules for use with meat, poultry, and fresh fruit.

The United States Department of Agriculture (USDA) has approved the use of low-level irradiation as an alternative treatment to pesticides for fruits and vegetables that are considered hosts to a number of insect pests, including fruit flies and seed weevils. Under bilateral agreements that allows less-developed countries to earn income through food exports agreements are made to allow them to irradiate fruits and vegetables at low doses to kill insects, so that the food can avoid quarantine.

European law dictates that all member countries must allow the sale of irradiated dried aromatic herbs, spices and vegetable seasonings. However, these Directives allow Member States to maintain previous clearances food categories the EC's Scientific Committee on Food (SCF) had previously approved (the approval body is now the European Food Safety Authority). Presently, Belgium, Czech Republic, France, Italy, Netherlands, Poland, and the United Kingdom allow the sale of many different types of irradiated foods. Before individual items in an approved class can be added to the approved list, studies into the toxicology of each of such food and for each of the proposed dose ranges are requested. It also states that irradiation shall not be used "as a substitute for hygiene or health practices or good manufacturing or agricultural practice". These Directives only control food irradiation for food retail and their conditions and controls are not applicable to the irradiation of food for patients requiring sterile diets.

Because of the Single Market of the EC any food, even if irradiated, must be allowed to be marketed in any other Member State even if a general ban of food irradiation prevails, under the condition that the food has been irradiated legally in the state of origin. Furthermore, imports into the EC are possible from third countries if the irradiation facility had been inspected and approved by the EC and the treatment is legal within the EC or some Member state.

Australia banned irradiated cat food after a national scare where cats suffered from paralyzation after eating a specific brand of highly irradiated catfood for an extended period of time. The suspected culprit was malnutrition from consuming food depleted of Vitamin A by the irradiation process. The incident was linked only to a single batch of one brand's product and no illness was linked to any of that brand's other irradiated batches of the same product or to any other brand of irradiated cat food. This, along with incomplete evidence indicating that the cat food was not sufficiently depleted of Vitamin A makes irradiation a less likely cause. Further research has been able to experimentally induce the paralyzation of cats by via Vitamin A deficiency by feeding highly irradiated food. For more details see the Long term impacts section.

Nuclear Safety and Security

Interlocks and safeguards are mandated to minimize this risk. There have been radiation-related accidents, deaths, and injury at such facilities, many of them caused by operators overriding the safety related interlocks. In a radiation processing facility, radiation specific concerns are supervised by special authorities, while "Ordinary" occupational safety regulations are handled much like other businesses.

The safety of irradiation facilities is regulated by the United Nations International Atomic Energy Agency and monitored by the different national Nuclear Regulatory Commissions. The regulators enforce a safety culture that mandates that all incidents that occur are documented and thoroughly analyzed to determine the cause and improvement potential. Such incidents are studied by personnel at multiple facilities, and improvements are mandated to retrofit existing facilities and future design.

In the US the Nuclear Regulatory Commission (NRC) regulates the safety of the processing facility, and the United States Department of Transportation (DOT) regulates the safe transport of the radioactive sources.

Irradiated Food Supply

As of 2010, the quantities of foods irradiated in Asia, the EU and the US were 285,200, 9,300, and 103,000 tons. Authorities in some countries use tests that can detect the irradiation of food items to enforce labeling standards and to bolster consumer confidence. The European Union monitors the market to determine the quantity of irradiated foods, if irradiated foods are labeled as irradiated, and if the irradiation is performed at approved facilities.

Irradiation of fruits and vegetables to prevent the spread of pest and diseases across borders has been increasing globally. In 2010, 18,446 tonnes of fruits and vegetables were irradiated in six countries for export quarantine control. 97% of this was exported to the United States.

In total, 103 000 tonnes of food products were irradiated on mainland United States in 2010. The three types of foods irradiated the most were spices (77.7%), fruits and vegetables (14.6%) and meat and poultry (7.77%). 17 953 tonnes of irradiated fruits and vegetables were exported to the mainland United States. Mexico, the United States' state of Hawaii, Thailand, Vietnam and India export irradiated produce to the mainland U.S. Mexico, followed by the United States' state of Hawaii, is the largest exporter of irradiated produce to the mainland U.S.

In total, 6 876 tonnes of food products were irradiated in European Union countries in 2013; mainly in four member state countries: Belgium (49.4%), the Netherlands (24.4%), Spain (12.7%)

and France (10.0%). The two types of foods irradiated the most were frog legs (46%), and dried herbs and spices (25%). There has been a decrease of 14% in the total quantity of products irradiated in the EU compared to the previous year 2012 (7 972 tonnes).

Pulsed Electric Field

Pulsed electric field (PEF) technology is a non-thermal food preservation method that involves the use of short electricity pulses for microbial inactivation while imposing minimal detrimental influence on food quality. This technology has the major advantage to provide high-quality foods to the consumers. PEF is claimed as superior to thermal processing and preservation methods because it reduces detrimental changes in food quality and nutrition and keeps physical and sensorial attributes of food.

PEF technology has wide range of applications ranging from liquid or semi-solid foods to solid foods. Most PEF studies have captivated on application of high voltage pulses for microbial inactivation milk and dairy products, egg products, juices and other liquid foods . However, most researchers have mainly focused on the aspect of food preservation with special reference to the microbial control and a lesser information is presented about the effect of PEF on food composition, quality and acceptability. Recently, some investigations have been conducted to evaluate the potential of PEF for improving food processing efficiency like enhancement of juice extraction and escalation of the food dehydration or drying.

Wouters et al. reported that PEF is known as one of the most auspicious non-thermal tools for microbial decontamination of foods. It involves the generation of electric fields(5-50 kV/cm) with the help of short high voltage pulses (us) between two electrodes that leads to microbial inactivation at temperatures lower than thermal methods. They also proposed that exact mechanistic approach of microbial inactivation is not known however, it is normally hypothesized that PEF causes penneabilization or depolarization of microbial membranes.

Studies revealed that PEF technology enables inactivation of bacterial and yeast vegetative cells in various foods. On the contrary, bacterial spores cannot be killed by employing pulsed electric fields because spores are resistant to PEF. Thus, applications of PEF are primarily focused on pathogenic and spoilage causing microorganisms in food. In addition to the pronounced effect of this technology in controlling microbiological spoilage of foods in a rapid and uniform manner, PEF also delivers shelf life extension without using heat treatment and preservation of sensorial and nutritional quality of foods. Likewise, PEF is also capable to improve the energy usage in an efficient and economical way. Hence, successful applications of PEF technology propose an alternative to conventional thermal processing techniques for food preservation and processing. In this topic, principles, mechanisms and recent applications of PEF technology are reviewed and compared with thermal processing technologies. Furthermore, examples are coded to illustrate the potential of PEF technology for aiming at preserving the quality features of various foods.

Principle of PEF

PEF technology involves the use of pulses having higher electric fields for only a few micro to milliseconds with intensity in the range of 10-80 kV/cm. The process depends on the number of pulses delivered to the product which is held between two electrodes. These electrodes have a specific gap between them which is known as treatment gap of the chamber. During PEF processing, high voltage is applied that results in the inactivation of microorganisms present in the food sample. The electric field is applied in different forms like as exponentially decaying waves, bipolar waves or oscillatory pulses. The process can also be carried at various temperature ranges such as ambient, sub-ambient and above-ambient temperatures. Food is packed after treatment with PEF and then stored under refrigerated conditions.

The science involved behind the transfer of electric pulses from food is that food contains several ions that provide a definite level of electrical conductivity to the product. This technique is usually preferred for liquid foods because electrical current flows into the liquid food more efficiently and the transfer of pulses from one point to other in liquids is quite easy due to the presence of charged molecules present.

A group of researchers, Zimmermann & Benz stated that mechanism of the functioning of PEF technology is the delivery of pulsing power to the product that is placed between a set of electrodes confining the treatment gap of the PEF chamber. The typical system has a pulse generator that produces high voltage pulses, treatment chamber that handles the product to be treated and associated with controlling and monitoring devices. Food product is placed in the treatment chamber equipped with electrodes connected with each other by a nonconductive material which prevents the flow of electricity from one to the other. High voltage electrical pulses are generated that are transferred to the product. The product placed between the electrodes experiences a force per unit charge which ruptures the bacterial cell membranes . In general, PEF technology is suggested for pasteurization of various food products including milk, juices, yogurt, liquid eggs and soups . Furthermore, combination of PEF with ultrasound, high pressure and ultraviolet light treatments may enhance the process output.

Microbial Inactivation by PEF

Several studies have investigated the mode of action of PEF to reduce the microbial load in various food products. Nonetheless, the exact mechanistic approach underlying the microbial inactivation by PEF has not been fully expounded as yet. However, a general mechanism of PEF action involves instability of microbial membranes by the induction of electrical field and electromechanical compression that leads to the pore formation in membrane . Mechanical fickleness of membranes is caused due to critical membrane potential which is formed by electrical field. Electroporation results in a significant increase in the membrane rupture and permeability which is termed as electro permeabilization. Electro permeabilization can be reversible or irreversible, depending on the degree of membrane organizational changes that results in cell death. Literature explains that membrane permeability is increased in a considerable manner by increasing the strength of electric field. This elevated membrane instability harmonizes with inactivation of microbial cells. In general, spores are claimed as more obstinate to PEF treatment as compared to vegetative cells.

Factors Influencing PEF Performance to Inactivate Microorganisms

The capability of PEF to inactivate microbes depends on several factors that can be categorized as process parameters, nature of product and properties of microbial cells . These factors play a dynamic role to attain the optimum performance of PEF treatment. An overview of these parameters is discussed herein.

Process Parameters

Several process parameters affect the ability of PEF to reduce microbial population in food such as strength of electric field, pulse length and shape, number of pulses and temperature. On a general node, increased intensity of these factors improves microbial inactivation but their exact relationship with the survival rate of microorganisms is not clear. Therefore, exact measurement of all these parameters is required to acquire reliable results.

Product Nature

The administration of PEF is also influenced by nature of product as a vast range of products are being treated by PEF which include fruit juices, liquid eggs, milk and dry herbs. It is investigated through experimental trials that PEF treatment is not much effective for products having particles or special structures, i.e. emulsions, Additionally, physical and chemical properties of the food also affect the rate of microbial decontamination. Various studies have revealed the influence of pH, water activity and electrical conductivity on the efficiency of PEF to inactivate microorganisms.

pH has a significant influence on the inactivation kinetics of microbes. Jeantet et al. reported that higher inactivation of Salmonella was observed in foods having neutral or above neutral pH values. Similarly, Alvarez et al. testified substantial reduction in L. mono cytogenes number in high acid foods such as citrus juices. Likewise, conductivity of the treatment medium has inverse relation with microbial inactivation. It is observed that foods having high electrical conductivity show less inactivation of microorganisms after PEF treatment. On the contrary, water activity has a direct relation with microbial reduction by PEF treatment as confirmed by Min & Zhang.

Characteristics of Microbes

Inactivation of microorganism by employing PEF technology also depends on microbial characteristics including type of microorganisms, species and strains. Generally, Gram- positive and Gram-negative bacteria are thought to be more resistant in comparison with yeast cells. Likewise, bacterial and mold spores are also claimed as recalcitrant to PEF treatment. Additionally, cell size and shape also affect the inactivation kinetics due to the difference in the development of critical membrane potential. PEF treatment affects different bacterial species at altered rate. It is usually proposed that Salmonella and E. coli are more susceptible to PEF as compared to Listeria and Bacillus species. Growth conditions like temperature, growth medium, concentration of nutrients and pH of the treated medium also influence PEF efficiency as well.

Applications in Food Industry

Application of PEF technology has been extensively demonstrated for the pasteurization of various food products like juices, milk & dairy products, soup and liquid eggs. However, it has several limitations such as product must be free from air bubbles and must have lower electrical conductivity. Additionally, particle size should be less than the gap of the treatment region to ensure appropriate treatment. PEF is generally not suitable for solid foods however several solid products have also been investigated to be efficiently treated by deploying PEF treatment. PEF technology can also be used to enhance the extraction of several bioactive components and sugars from plant cells .

PEF processing has shown its potential to treat less viscous fruit juices having less electrical conductivity such as apple, citrus and cranberry juices. PEF technology also executes beneficial aspects on the quality parameters of fruit juice. Correspondingly, Yeom et al. compared pasteurized and PEF-treated citrus juice during refrigerated storage (4°C) for a period of 112 days and observed less browning in PEF-processed juice comparatively to traditionally pasteurized juice due to conversion of ascorbic acid to furfural. Besides, PEF-treated foods also retained their fresh flavor, textural & functional attributes and extended shelf life in addition with microbiological safety . In recent years, PEF technology has been utilized for various purposes like enhancement of drying efficiency, modification of enzymatic activity, solid food preservation, waste water treatment and extraction.

PEF in Fruit Processing

PEF processing has promising applications in citrus industry with special reference to the inactivation of microorganisms and prevention of developing off-flavors during the storage . Jemai & Vorobiev noticed that treatment of apple juice with PEF resulted in enhancement of diffusion coefficients of soluble substances. This technology can also be fruitfully applied for the disintegration of biological tissues that enhances the extraction of intracellular compounds from different fruits. For instance, pectin is a very useful component found in many fruits that is traditionally being extracted through enzymatic reaction but this reaction has less yield of pectin due to poor efficiency. Alternatively, PEF treatment is employed as short pulses to avoid excessive heat and undesirable electrolytic reactions that can enhance the extraction rate of pectin from fruit pomace.

Bacterial Inactivation in Milk by PEF

Inadequately pasteurized milk may cause several health hazards due to the presence of several spoilage causing and pathogenic bacteria particularly Escherichia coli and Listeria and Pseudomonas spp . Elevated concerns regarding the impact of heat treatment on the quality of milk and consumer demand for fresh-like quality attribute products have encouraged the induction of PEF for milk pasteurization. PEF-induced microbial inactivation is believed to be an effective way of milk preservation without adversely affecting the milk quality.

Studies have demonstrated the effectiveness of PEF processing for microbial reduction in simulated milk ultra-filtrate and skim milk . However, presence of fat and protein moieties limits the adeptness of PEF in whole milk because these molecules protect bacterial cells during treatment . Therefore, it is imperative to validate the worth of PEF treatment to inactivate bacteria in a complex whole milk matrix for a genuine comparison with thermal pasteurization.

Garcia et al. observed sub-lethal injury in bacterial cells and concluded that PEF treatment can be successfully used in synergy with other hurdles to get more benefits. In another trial, PEF processing of milk was combined with heat treatment up to 55-60°C and a significant reduction was observed in the microbial load . More recently, Sharma et al. employed PEF treatment at different dose, time and temperature ranges to inactivate Gram-positive and Gram-negative bacteria in whole milk and reported 5-6 log reduction in bacterial number at 22-28 kV cm-1 for 17-101 μs at 50°C. On the other hand, some studies did not clearly demonstrate the effects pre-heating, temperature, time & dose variation on PEF efficiency and effects of PEF treatment on milk quality and functional properties. Therefore, it is questionable whether PEF can maintain the integrity of heat-sensitive milk components while inactivating spoilage and pathogenic microorganisms in whole milk and need further exploration.

PEF and Meat Quality

Meat products have been widely consumed around the globe due to the presence of valuable micronutrients high quality protein . Quality of meat is at prime importance because meat quality is considered as the most vital factor for purchasing decisions of consumers . Pulsed electric field technology has shown its potential for various applications in solid foods with different aims including structural modifications, changing physical quality parameters, extraction of bioactive compounds and preservation. Nevertheless, this technology has limited applications in muscle foods . PEF technology can be employed in meat processing for various purposes including enhancement of cell permeation to increase tenderness, attenuation of microbial load to improve the shelf life and maintenance of volatile profile of meat during storage .

PEF technology can considerably improve enzyme release and glycolysis that are essential for proteolysis for meat tenderization. O'Dowd et al. studied the effect of PEF on meat quality @ 1.1-2.8 kV cm−1 and reported no significant improvement in meat tenderness during storage. Later on, Bekhit et al. found a weighty improvement in beef tenderness by PEF treatment (5-10 kV). Likewise, applied PEF (5-10 kV) at different frequencies (20, 50 and 90 Hz) on beef muscles and revealed that PEF treatment reduced shear force up to 19% & improved tenderness by augmenting the degradation of desmin and troponin-T during a refrigeration storage of 21 days.

Modified Atmosphere

Modified Atmosphere Packaging is also known as gas flushing, protective atmosphere packaging or reduced oxygen packaging.

Modified Atmosphere enables fresh and minimally processed packaged food products to maintain visual, textural and nutritional appeal. The controlled MAP environment enables food packaging to provide an extended shelf life without requiring the addition of chemical preservatives or stabilisers. Processors and marketers of food products rely on Modified Atmosphere Packaging to assure fresh and flavourful products that continually meet the consumer's expectation for brand quality, consistency, freshness and in-stock availability.

Modified Atmosphere Packaging is an optimal blend of pure oxygen, carbon dioxide and nitrogen within a high barrier or permeable package. A finely adjusted and carefully controlled gas blend is developed to meet the specific respiration needs for each packaged food product.

Plastic films, foils and other packaging materials that demonstrate specified gas permeability properties and/or water vapour permeability properties are selected for use. These high barrier substrates become MAP Packages after they are formed into trays, lid stock or bags and filled with a select blend of oxygen, carbon dioxide and nitrogen environmental gasses.

Packaging films are selected to match the characteristics and needs of the food product. Film permeability, water vapour transmission rates and sealing characteristics need to be measured and tested at film selection and again at package converting and product fill stages, since the ability of a film to handle MAP performance characteristics may vary within each stage.

Working of Modified Atmosphere Packaging

The Modified Atmosphere Package environment is formed from a finely balanced mix of normal atmospheric gases. The finely balanced MAP gas mix slows down the product aging process to reduce colour loss, odour and off-taste resulting from product deterioration, spoilage and rancidity caused by mold and other anaerobic organisms.

A carefully controlled Modified Atmosphere Package achieves and maintains an optimal respiration rate to preserve the fresh colour, taste and nutrient content of red meat, seafood, minimally processed fruits and vegetables, pasta, prepared foods, cheese, baked goods, cured meats and dried foods throughout an extended shelf life.

Modified Atmosphere Packaging Offers Supply Chain Efficiencies

Longer shelf life MAP packages allow food processors, food manufacturers, food distributors and food retailers to better control product quality, availability and costs.

Longer freshness cycles permit grocers to eliminate frequent product rotation, removal and restocking; thereby reducing labour and waste disposal costs.

Distributors can extend distribution territories or offer a greater variety of product lines to the retailer, since less frequent product replacement requirements permits growth in other areas.

Food manufacturers are able to take advantage of extended replacement cycles to reduce production replacement demands. Manufacturing capacity can be more profitably utilised by developing and offering new products.

The first recorded beneficial effects of using modified atmosphere date back to 1821. Jacques Etienne Berard, a professor at the School of Pharmacy in Montpellier, France, reported delayed ripening of fruit and increased shelf life in low-oxygen storage conditions. Controlled Atmosphere Storage (CAS) was used from the 1930's when ships transporting fresh apples and pears had high levels of CO_2 in their holding rooms in order to increase the shelf life of the product. In the 1970s MA packages reached the stores when bacon and fish were sold in retail packs in Mexico. Since then development has been continuous and interest in MAP has grown due to consumer demand.

Theory

Atmosphere within the package can be modified passively or actively. In passive modified atmosphere packaging (MAP), the high concentration of CO_2 and low O_2 levels in the package is achieved over time as a result of respiration of the product and gas transmission rates of the packaging film. This method is commonly used for fresh respiring fruits and vegetables. Reducing O_2 and increasing CO_2 slows down respiration rate, conserves stored energy, and therefore extended shelf life. On the other hand, active MA involves the use of active systems such as O_2 and CO_2 scavengers or emitters, moisture absorbers, ethylene scavengers, ethanol emitters and gas flushing in the packaging film or container to modify the atmosphere within the package.

The mixture of gases selected for a MAP package depends on the type of product, the packaging materials and the storage temperature. The atmosphere in an MA package consists mainly of adjusted amounts of N_2, O_2, and CO_2. Reduction of O_2 promotes delay in deteriorative reactions in foods such as lipid oxidation, browning reactions and growth of spoilage organisms. Low O_2 levels of 3-5% are used to slow down respiration rate in fruits and vegetables. In the case of red meat, however, high levels of O_2 (~80%) are used to reduce oxidation of myoglobin and maintain an attractive bright red color of the meat. Meat color enhancement is not required for pork, poultry and cooked meats; therefore, a higher concentration of CO_2 is used to extend the shelf life. Levels higher than 10% of CO_2 are phytotoxic for fruit and vegetables, so CO_2 is maintained below this level. N_2 is mostly used as a filler gas to prevent pack collapse. In addition, it is also used to prevent oxidative rancidity in packaged products such as snack foods by displacing atmospheric air, especially oxygen, therefore extending shelf life. The use of noble gases such as Helium (He), Argon (Ar) and Xenon (Xe) to replace N_2 as the balancing gas in MAP can also be used to preserve and extend the shelf life of fresh and minimally processed fruits and vegetables. Their beneficial effects are due to their higher solubility and diffusivity in water, making them more effective in displacing O_2 from cellular sites and enzymatic O_2 receptors.

There has been a debate regarding the use of carbon monoxide (CO) in the packaging of red meat due to its possible toxic effect on packaging workers. Its use results in a more stable red color of carboxymyoglobin in meat, which leads to another concern that it can mask evidence of spoilage in the product.

Effect on Microorganisms

Low O_2 and high CO_2 concentrations in packages are effective in limiting the growth of Gram negative bacteria, molds and aerobic microorganisms, such as *Pseudomonas* spp. High O_2 combined with high CO_2 could have bacteriostatic and bactericidal effects by suppression of aerobes by high CO_2 and anaerobes by high O_2. CO_2 has the ability to penetrate bacterial membrane and affect intracellular pH. Therefore, lag phase and generation time of spoilage microorganisms are increased resulting in shelf life extension of refrigerated foods. Since the growth of spoilage microorganisms are suppressed by MAP, the ability of the pathogens to grow is potentially increased. Microorganisms that can survive under low oxygen environment such as *Campylobacter jejuni*, *Clostridium botulinum*, *E. coli*, *Salmonella*, *Listeria* and *Aeromonas hydophila* are of major concern for MA packaged products. Products may appear organoleptically acceptable due to the delayed growth of the spoilage microorganisms but might contain harmful pathogens. This risk can be minimized by use of additional hurdles such as temperature control (maintain temperature below 3 degrees C), lowering water activity (less than 0.92), reducing pH (below 4.5) or addition or preservatives such as nitrite to delay metabolic activity and growth of pathogens.

Packaging Materials

Flexible films are commonly used for products such as fresh produce, meats, fish and bread seeing as they provide suitable permeability for gases and water vapor to reach the desired atmosphere. Pre-formed trays are formed and sent to the food packaging facility where they are filled. The package headspace then undergoes modification and sealing. Pre-formed trays are usually more flexible and allow for a broader range of sizes as opposed to thermoformed packaging materials as different tray sizes and colors can be handled without the risk of damaging the package. Thermoformed packaging however is received in the food packaging facility as a roll of sheets. Each sheet is subjected to heat and pressure, and is formed at the packaging station. Following the forming, the package is filled with the product, and then sealed. The advantages that thermoformed packaging materials have over pre-formed trays are mainly cost-related: thermoformed packaging uses 30% to 50% less material, and they are transported as rolls of material. This will amount in significant reduction of manufacturing and transportation costs.

When selecting packaging films for MAP of fruits and vegetables the main characteristics to consider are gas permeability, water vapour transmission rate, mechanical properties, transparency, type of package and sealing reliability. Traditionally used packaging films like LDPE (low-density polyethylene), PVC (polyvinyl chloride), EVA (ethylene-vinyl acetate) and OPP (oriented polypropylene) are not permeable enough for highly respiring products like fresh-cut produce, mushrooms and broccoli. As fruits and vegetables are respiring products, there is a need to transmit gases through the film. Films designed with these properties are called permeable films. Other films, called barrier films, are designed to prevent the exchange of gases and are mainly used with non-respiring products like meat and fish.

MAP films developed to control the humidity level as well as the gas composition in the sealed package are beneficial for the prolonged storage of fresh fruits, vegetables and herbs that are sensitive to moisture. These films are commonly referred to as modified atmosphere/modified humidity packaging (MA/MH) films.

Equipment

In using form-fill-seal packaging machines, the main function is to place the product in a flexible pouch suitable for the desired characteristics of the final product. These pouches can either be pre-formed or thermoformed. The food is introduced into the pouch, the composition of the headspace atmosphere is changed within the package; it is then heat sealed. These types of machines are typically called pillow-wrap, which horizontally or vertically form, fill and seal the product. Form-fill-seal packaging machines are usually used for large scale operations.

In contrast, chamber machines are used for batch processes. A filled pre-formed wrap is filled with the product and introduced into a cavity. The cavity is closed and vacuum is then pulled on the chamber and the modified atmosphere is inserted as desired. Sealing of the package is done through heated sealing bars, and the product is then removed. This batch process is labor intensive and thus requires a longer period of time; however, it is relatively cheaper than packaging machines which are automated.

Additionally, snorkel machines are used to modify the atmosphere within a package after the food has been filled. The product is placed in the packaging material and positioned into the machine without the need of a chamber. A nozzle, which is the snorkel, is then inserted into the packaging material. It pulls a vacuum and then flushes the modified atmosphere into the package. The nozzle is removed and the package is heat sealed. This method is suitable for bulk and large operations.

Products

Many products such as red meat, seafood, minimally processed fruits and vegetables, salads, pasta, cheese, bakery goods, poultry, cooked and cured meats, ready meals and dried foods are packaged under MA. A summary of optimal gas mixtures for MA products are shown in table.

Modified Atmosphere Packaging for different food products and optimal gas mixtures

Product	Oxygen (%)	Carbon Dioxide (%)	Nitrogen (%)
Red Meat	80 - 85	15	-
Poultry	25	75	-
Fish	-	60	40
Soft Cheeses	-	100	-
Hard Cheeses	-	100	-
Bread	70	30	-
Fresh Pasta	-	-	100
Fruits and Vegetables	3 - 5	3 - 5	85 – 95

Nonthermal Plasma

Processed foods such as nutritional losses and adverse effects on organoleptic quality. This has led to the emergence of so called "Nonthermal Technologies". Nonthermal technologies are preservation treatments that are effective at ambient or sub-lethal temperatures, thereby minimizing negative thermal effects on nutritional and quality parameters of food. These include the application of gamma irradiation, beta irradiation (electron beam), power ultrasound, ozonation, pulsed light, UV treatment, pulsed electric field (PEF), high hydrostatic pressure etc. Nevertheless, even some of these commercially viable inactivation methods are limited in practice due to adverse perceptions associated (like with treatments of gamma irradiation and high energy electron beams) or high initial investments required and/or other constraints. Purely physical techniques, such as high hydrostatic pressure, are chemically safe but require complex or expensive equipment, affect the quality of the product (Kruk et al. 2009) and are generally incompatible with online treatments. Very few approaches are suitable for treatment of solid foods, in particular fruit and vegetables. Very recently NTP has been added to the existing list of non-thermal processes for the decontamination of fresh produce.

Technologies like UV treatment, zonation, power ultrasound, pulsed light, electric discharge and non thermal plasma are commonly designated as Advanced Oxidation Processes (AOP). The use of pulsed UV light as a means of microbial inactivation is a mature technology that has commercial application in surface disinfection of packaging materials, but demonstrates limitations due to shadowing effects in food products. Indeed, there is currently no perfect method to achieve sterilization at ambient temperature.

High-pressure Food Preservation

High pressure processing is a promising "non-thermal" technique for food preservation that efficiently inactivates the vegetative microorganisms, most commonly related to food- borne diseases. High pressure processing is carried out with intense pressure in the range of 100-1000 MPa, with or without heat, allowing most foods to be preserved with minimal effect on taste, texture or nutritional characteristics. The main advantage of high pressure processing compared to thermal sterilization and pasteurization is maintenance of sensory and nutritional characteristic of treated food products.

Pressure treatment can be used to process both liquid and high-moisture-content solid foods. Pressure processing is a lethal to microorganisms but at relatively low temperatures (0-40°c) covalent bonds are almost unaffected. The limited effect of hPP (at moderate temperature) on covalent bonds represents a unique characteristic of this technology because hPP has a minimal effect on food chemistry. hPP provides a means for retaining food quality while avoiding the need for excessive thermal treatments or chemical preservation.

Microbial inactivation is one of the main goals for the application of high pressure technology. the inactivation effect of high pressure processing results in extending shelf- life and improving the microbial safety of food products. According to some researches high pressure treatment could be accepted as a food safety intervention for eliminating Listeria monocytogenes in processed meat products and cheese. Hydrostatic pressure treatment is also effective in inactivating other hazardous microorganisms such as E. coli, Salmonella, and Vibrio, as well as many yeasts, molds, and bacteria responsible for food spoilage. The microbiological shelf-life and food quality can be substantially extended by the use of HPP.

Bacterial spores represent a challenge for high pressure technology and more information about their resistance is required. Microbial spores suspended in foods or laboratory model system could be inactivated by high pressure treatment but compared to vegetative cells the treatment conditions must be extreme: higher pressure and long exposure time at elevated temperature. When pressure-temperature is combined at 690 MPa and 80°c for 20 minutes, the treatments was effective with a significant reduction in the Clostridium sporogenes spore count. Successful treatment of Bacillus stearothermophilus is observed were pressure treatments is combined with moderate temperature (70°C). There are no published reports on the high-pressure resistance of Clostridium botulinum spores, and their ability to withstand high pressure at low or high temperatures is unknown.

Pressure treatment can also be used to alter the functional and sensory properties of various food components, especially proteins. the tertiary and quaternary structures of molecules which are maintained mainly by hydrophobic and ionic interactions are beneficially altered by high pressure above 200 MPa. the hydrophobic and electrostatic interactions are most affected but not the hydrogen bonds which stabilize α-helical and ß-pleated sheets. Meat, fish, egg and dairy proteins can be denatured with hPP in the absence of elevated temperatures. increased viscosity and opacity are obtained with little change in fresh flavour. On the other hand, high pressure has very little effect on low-molecular-weight compounds such as flavour compounds, vitamins, and pigments compared to thermal processes. Accordingly, the quality of hPP pasteurized food is very similar to that of fresh food products. the quality throughout shelf-life is influenced more by subsequent distribution and storage temperatures and the barrier properties of the packaging rather than by the high pressure treatments.

Basic High Pressure Processing Principles

High-Pressure technology has been cited as one of the best innovations in food processing from the last 50 years. Some physical and chemical changes result from application of pressure. Physical compression during pressure treatment results in a volume reduction and an increase in temperature and energy.

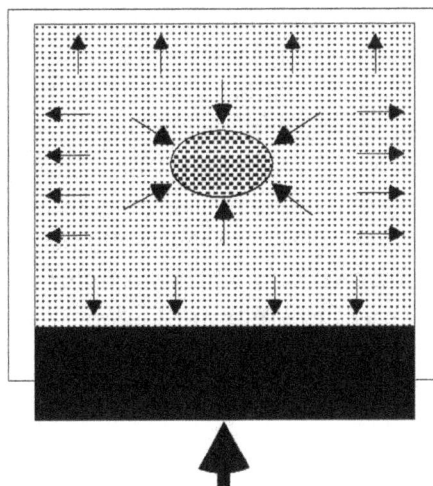

The principle of isostatic processing

The basic principles that determine the behavior of foods under pressure are:

- Le Chatelier's principle: any reaction, conformational change, phase transition, accompanied by a decrease in volume is enhanced by pressure;

- Principle of microscopic ordering: at constant temperature, an increase in pressure increases the degrees of ordering of molecules of a given substance. therefore pressure and temperature exert antagonistic forces on molecular structure and chemical reactions;

- Isostatic principle: the principle of isostatic processing is presented in the food products are compressed by uniform pressure from every direction and then returned to their original shape when the pressure is released. the products are compressed independently of the product size and geometry because transmission of pressure to the core is not mass/time dependant thus the process is minimized.

If a food product contains sufficient moisture, pressure will not damage the product at the macroscopic levels as long as the pressure is applied uniformly in all directions.

High Pressure Equipment

Even if the principles of HPP and its influence on microbial inactivation have been recognized since late 1800's, the first commercial installation for HPP appeared in Japan in 1990. Although high pressure technology is currently more expensive than traditional processing technologies, the use of high pressure offers new opportunities for food industry to respond to the demand from consumers.

A high-pressure system consists of a high-pressure vessel and its closure(s) pressure-generation system, temperature- control device and material-handling system. the pressure vessel is the most important component of high- hydrostatic-pressure equipment. Several aspects must be taken into account in vessel design. it is necessary to design the high-pressure vessel to be dimensionally stable in a safe- fail way. if it fails it should fail with leak before fracture. Pressure-transmitting fluids are used in the vessel to transmit pressure uniformly and instantaneously to the products sample. Most widely used fluids are water, glycol solutions, silicone oil, sodium benzoate solutions, ethanol solutions, inert gases and castor oil. the food products should be packaged in a flexible packaging. The packages are loaded into the high pressure chamber. the vessel is sealed and the vessel filled with pressure transmitting agent. The high pressure is usually carried out with water as a hydraulic fluid to facilitate the operation and compatibility with food materials. the basis for applying high pressure to foods is to compress the water surrounding the food. At room temperature, the volume of water decreases with an increase in pressure. Because liquid compression results in a small volume change, high-pressure vessels using water do not present the same operating hazards as vessels using compressed gases. ones the desired pressure is reached the pump or piston is stopped, the valves are closed and the pressure can be maintained without further energy input. After holding the product for the desired time at the target pressure, the vessel is decompressed by releasing the pressure-transmitting fluid. For most applications, products are held for 3-5 min at 600 MPa. Approximately 5-6 cycles per hour are possible, allowing time for compression, holding, de-compression, loading and unloading. After pressure treatment, the processed product is removed from the vessel and stored in a conventional way.

High pressures can be generated by direct or indirect compression or by heating the pressure fluid.

Direct Compression

It is generated by pressurizing a fluid by a piston, driven at its larger diameter end by a low pressure pump. This method allows very fast compression, but the limitations of the high-pressure dynamic seal between the piston and the vessel's internal surface restrict the use of this method to small-diameter laboratory or pilot plant systems.

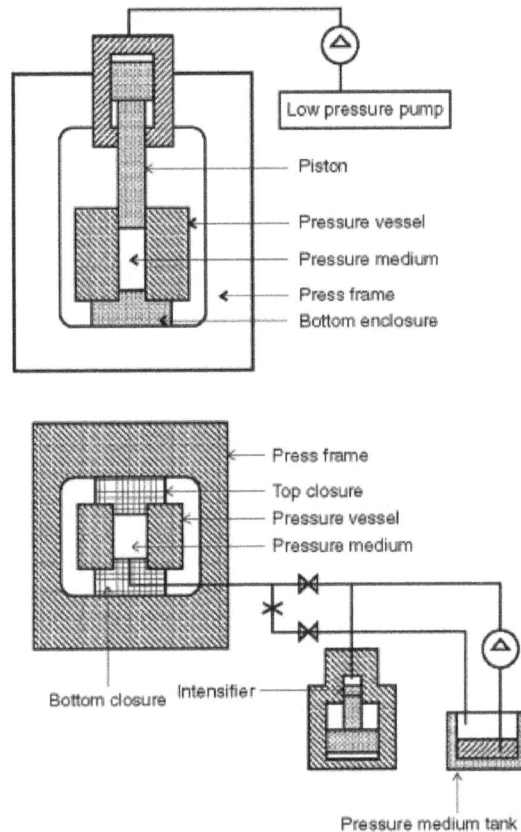

Generation of high pressure by direct (top) and indirect (bottom) compression of the pressure-transmitting medium

Indirect Compression

This technique uses a high-pressure intensifier to pump a pressure medium from a reservoir into a closed high-pressure vessel until the desired pressure is reached.

Heating of the Pressure Medium

It utilizes expansion of the pressure fluid with increasing temperature to generate high pressure. this method is therefore used when high pressure is applied in combination with high temperature and requires very accurate temperature control within the entire internal volume of the pressure vessel.

Pressure-temperature Effect

to understand and foresee the effect of HPP on foods it is necessary to take in attention the net combined pressure- temperature effect on the treated foods.

During the compression phase $(t_1 - t_2)$ of pressure treatment food products undergo a decrease in volume as a function of the pressure. the product is held under pressure for a certain time $(t_2 - t_3)$ before decompression $(t_3 - t_4)$. Upon decompression, the product will usually expand back to its initial volume. The compression and decompression can result in a transient temperature change in the product during treatment. The temperature of food $(T_1 - T_2)$ increases as a result of physical compression $(P_1 - P_2)$. Product temperature $(T_2 - T_3)$ at process pressure $(P_2 - P_3)$ is independent of compression rate as long as heat exchange between the product and the surroundings is negligible. In a perfectly insulated (adiabatic) system, the product will return to its initial temperature upon decompression $(P_3 - P_4)$. In practice, however, the product will return to a temperature (T_4) slightly lower than its initial temperature (T_1) as a result of heat losses during the compression phase. The rapid heating and cooling resulting from HPP treatment offer a unique way to increase the temperature of the product only during the treatment and to cool it rapidly thereafter. The temperature of water increases about 3°C for every 100 MPa of increased pressure at room temperature. On the other hand, fats and oils have a heat of compression value of 8-9°C/100 MPa, and proteins and carbohydrates have intermediate heat of compression values (41, 44).

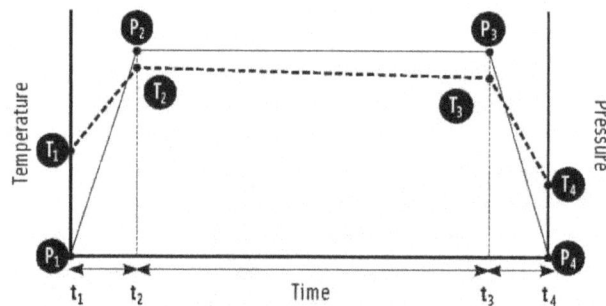

Pressure-temperature history during high pressure processing

High Pressure Processing and Microbial Inactivation

Microbial inactivation is one of the main goals for the application of high pressure technology.

The extent of microbial inactivation that is achieved by suitable high pressure treatment depends on a number of interacting factors, including type and number of microorganisms, magnitude and duration of high pressure treatments, temperature and composition of the suspension media or food.

The pressure sensitivity of microorganisms may vary between species and probably among the strains of the same species. Microorganisms can be divided into those that are relatively pressure sensitive and those that are pressure resistant. Generally Gram-positive bacteria are more resistant to pressure than Gram-negative bacteria, moulds and yeasts. Most resistant to high hydrostatic pressure are bacterial spores. the sensitivity of microbial cells depends on the stage of the growth cycle at which the organisms are subjected to high hydrostatic pressure treatment. in general, cells in the exponential phase are more sensitive to pressure treatments than cells in the log or stationary phases of growth (16, 42). the grater resistance to pressure when the cell metabolism is slowed down may be due to accumulation of cell components that reduce the effect of high pressure. Rich nutrient media such as meat reinforce the resistance of the microorganisms to HPP. Carbohydrates, proteins and lipids also have a protective effect. A low water activity protects microorganisms against pressure and tends to inhibit pressure inactivation with noticeable retardation as water activity falls.

The extent and duration of high pressure treatment influence the microbial inactivation. An increase in pressure increases microbial inactivation. however, the duration of treatment is increased that does not necessary leads to an increase in the lethal effect. As mentioned above, the microbial response to high-pressure treatments depends on the type of microorganism. For each of them, there is a pressure-level threshold beyond which no effects are detected by increasing the exposure time. There also exists a pressure level at which increasing the treatment time causes significant reductions in the initially inoculated microbial counts.

The temperature during high pressure treatment influences the range of inactivation of microbial cells. Some authors showed that the pressure resistance of microbial organisms is maximal at temperature 15-30°C and decreases significantly at higher or lower temperature.

The treatment of microbial cell and spores with high pressure results in many changes in the morphology, cell membranes, biochemical aspects and genetic mechanisms and all these processes are related to the inactivation of microorganisms. the lethal effect of high pressure on vegetative microorganisms is thought to be the result of a number of possible changes that take place simultaneously in the microbial cell.

The membrane is the most probable site of disruption. high-hydrostatic pressure treatments can alter membrane functionalities such as active transport or passive permeability and therefore perturb the physicochemical balance of the cell. there is a considerably evidence that pressure tends to loosen the contact between attached enzymes and membrane surface as a consequence of the changes in the physical state of lipids that control enzyme activity. the leakage of intracellular constituent through the permeabilized cell membrane is the most direct reason for cell death after high pressure treatment.

Inactivation of key enzymes, including those involved in DNA replication and transcription is also mentioned as a possible inactivating mechanism.

Bacterial spores have demonstrated pressure resistance and the mechanisms through which they are inactivated are different from these for the vegetative cells. it has been suggested that the spore proteins are protected against solvation and ionization. Microbial spores could be inactivated by chosen suitable conditions for high pressure treatment: higher pressure and long exposure time at elevated temperature. it was assumed that pressure caused inactivation of spores by first initiating germination and then inactivating germinated forms. the spore germination could be induced by hydrostatic pressure of 100-300 MPa and resultant vegetative cells are sensitive to environmental conditions.

Usually for pasteurization purpose the considered treatment is generally in the range of 300-600 MPa for a short period of time, from seconds to minutes, inactivating vegetative pathogenic and spoilage microorganisms. For sterilization the range is over 600 MPa and combination with high temperature is needed because some spores are resistant even to pressure over than 1000 MPa when the temperature is not higher than 45-75°C. Most yeast are inactivated by exposure to 300-400 MPa at 25°C within a few minutes, however, yeast ascospores may require treatment at higher pressure. Pressure inactivation of moulds follows a model similar to yeast.

To explain the response of microorganisms to different pressure, high-pressure effects on several biological molecules have been studied. Protein denaturation, lipid phase change and enzyme inactivation can perturb the cell morphology, genetic mechanisms, and biochemical reactions. however, the mechanisms that damage the cells are still not fully understood.

Aspects of Applications of High Pressure Processing of Foods

High-pressure processing provides a unique opportunity for food processors to develop a new generation of value-added food products having superior quality and shelf-life to those produced conventionally.

High pressure processing is a very promising technology for ready-to-eat (Rte) meats because there are few barriers to approval by regulatory authorities, no special labelling requirements because no chemicals are added, and if used appropriately there are no changes to texture or flavour of the product. Researchers found that in Rte meats that are pressure treated at 600 MPa at 20°c for 180 sec, there were no changes in sensory quality, no difference in consumer acceptability, a 4 log reduction in Listeria monocytogenes in inoculated product and the refrigerated shelf-life was extended. There is report that hPP treatment (600 MPa for 10 minutes at 30°c) could extend the shelf-life of food including cooked ham, dry cured ham and marinated beef loins. high pressure application of 500 MPa could extends the shelf life of cooked pork ham and raw smoked pork loin up to 8 weeks, ensuring good microbiological and sensory quality of products .

High pressure processing can effectively inactivate the spoilage microorganisms of several foods, and important food- borne pathogens such as Campylobacter jejuni, Escherichia coli 0157:h7, Listeria monocytogenes, and Salmonella spp. Foods can be pasteurized at low or moderate temperatures under pressure. Pressurization at high temperature can sterilize foods. Pressure treatment is of special interest for products or meals containing ingredients that are extensively modified by heat.

HPP has potential as a phyto sanitary treatment to control quarantine insect pests in fresh or minimal processed fruits and vegetables to extend their shelf-life (7, 38). Pressure inactivation of yeast and moulds has been reported in citrus juices. Juices pressurized at 400 MPa for 10 min at 40°c did not spoil during 2-3 months of storage. the high pressure treatment effectively reduced the bacterial flora of fresh goat milk cheese and significantly extended the refrigeration storage life no surviving E. coli was detected in cheese after 60 days of storage (2-4°C) in inoculation studies after treatments at 400-500 MPa for 5-10 min.

High-pressure technology could improve the microbiological safety and quality of foods, including meat, milk, and dairy products.

Biopreservation

Biopreservation can be defined as the extension of shelf life and food safety by the use of natural or controlled micro biota and/or their antimicrobial compounds. One of the most common forms of food bio preservation is fermentation, a process based on the growth of microorganisms in foods, whether natural or added. These organisms mainly comprise lactic acid bacteria, which produce organic acids and other compounds that, in addition to antimicrobial properties, also confer unique flavours and textures to food products. Traditionally, a great number of foods have been protected against spoiling by natural processes of fermentation. Currently, fermented foods are increasing in popularity (60% of the diet in industrialized countries) and, to assure the homogeneity, quality, and safety of products, they are produced by the intentional application in raw foods of different microbial systems (starter/protective cultures). Moreover, because of the improved organoleptic qualities of traditional fermented food, extensive research on its microbial biodiversity has been carried out with the goal of reproducing these qualities, which are attributed to native micro biota, in a controlled environment.

The starter cultures of fermented foods can be defined as preparations of one or several systems of microorganisms that are applied to initiate the process of fermentation during food manufacture fundamentally in the dairy industry and, currently, extended to other fermented foods such as meat, spirits, vegetable products, and juices. The bacteria used are selected depending on food type with the aim of positively affecting the physical, chemical, and biological composition of foods, providing attractive flavour properties for the consumer. To be used as starter cultures, microorganisms must fulfill the standards of GRAS status (Generally Recognized as Safe by people and the scientific community) and present no pathogenic nor toxigenic potential. In addition, use must be standardized and reproducible. The same cultures have been employed for different uses and under different conditions. For the starter cultures, generally LAB, metabolic activity, such as acid production in cheese, is of great technological importance, whereas antimicrobial activity is secondary. However, for the protective culture, generally LAB also, the objectives are the opposite and must always take into account an additional factor for safety as its implantation must reduce the risk of growth and survival of pathogenic microorganism. An ideal strain would fulfil both the metabolic and antimicrobial traits.

Lactic Acid Bacteria

LAB include the genera Lactococcus, Streptococcus, Lactobacillus, Pediococcus, Leuconostoc, Enterococcus, Carnobacterium, Aerococcus, Oenococcus, Tetragenococcus, Vagococcus, and Weisella They form a natural group of Gram-positive, nonmotile, non-sporeforming, rod- and coccus-shaped organisms that can ferment carbohydrates to form chiefly lactic acid; they also have low proportions of G+C in their DNA (< 55%). LAB present attractive physiological properties and technological applications (resistance to bacteriophages, proteolytic activity, lactose and citrate fermentation, production of polysaccharides, high resistance to freezing and lyophilization, capacity for adhesion and colonization of the digestive mucosa, and production of antimicrobial substances).

In general, LAB have GRAS status and play an essential role in food fermentation given that a wide variety of strains are employed as starter cultures (or protective cultures) in the manufacture of dairy, meat, and vegetable products. The most important contribution of these microorganisms is the preservation of the nutritional qualities of the raw material through extended shelf life and the inhibition of spoilage and pathogenic bacteria. This contribution is due to competition for nutrients and the presence of inhibitor agents produced, including organic acids, hydrogen peroxide, and bacteriocins There are many reviews on reported examples of spoilage and pathogenic bacteria inhibition by bacteriocin-producing LAB.

In addition to the food applications of LAB, various strains are considered to be probiotics. Probiotics can be described as a preparation of or a product containing viable, defined microorganisms in sufficient numbers to alter the microbiota (by implantation or colonization) in a compartment of the host and that exert beneficial health effects in this host. In this regard, LAB fit many of requirements for a microorganism to be defined as an effective probiotic. These requirements include the ability to: (a) adhere to cells; (b) exclude or reduce pathogenic adherence; (c) persist and multiply; (d) produce acids, hydrogen peroxide, and bacteriocins antagonistic to pathogen growth; (e) be safe, noninvasive, noncarcinogenic, and nonpathogenic; and (f) coaggregate to form a normal balanced flora. Strains that are used as probiotics for man have been isolated from the human gastrointestinal tract and usually belong to species of the genera Lactobacillus and Bifidobacterium. However, strains belonging to species of other LAB have been used in the past as probiotics as well, such as E. faecium, E. faecalis, S. thermophilus, L. lactis subsp. lactis, Le. mesenteroides, and P. acidilactici.

LAB Bacteriocins

The antimicrobial ribosomally synthesized peptides produced by bacteria, including members of the LAB, are called bacteriocins. Such peptides are produced by many, if not all, bacterial species and kill closely related microorganisms. Due to their nature, they are inactivated by proteases in the gastrointestinal tract. Most of the LAB bacteriocins identified so far are thermostable cationic molecules that have up to 60 amino acid residues and hydrophobic patches. Electrostatic interactions with negatively charged phosphate groups on target cell membranes are thought to contribute to the initial binding, forming pores and killing the cells after causing lethal damage and autolysin activation to digest the cellular wall.

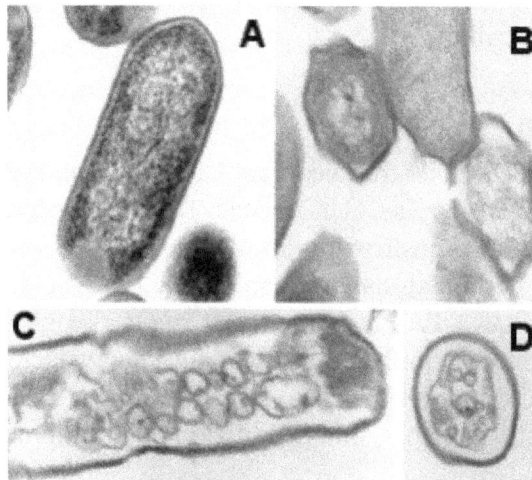

Example of damage caused by bacteriocin on L. monocytogenes CECT 4032 cells.

The LAB bacteriocins have many attractive characteristics that make them suitable candidates for use as food preservatives, such as:

- Protein nature, inactivation by proteolytic enzymes of gastrointestinal tract;
- Non-toxic to laboratory animals tested and generally non-immunogenic;
- Inactive against eukaryotic cells;

- Generally thermoresistant (can maintain antimicrobial activity after pasteurization and sterilization);

- Broad bactericidal activity affecting most of the Gram-positive bacteria and some, damaged, Gram-negative bacteria including various pathogens such as L. monocytogenes, Bacillus cereus, S. aureus, and Salmonella;

- Genetic determinants generally located in plasmid, which facilitates genetic manipulation to increase the variety of natural peptide analogues with desirable characteristics.

For these reasons, the use of bacteriocins has, in recent years, attracted considerable interest for use as biopreservatives in food, which has led to the discovery of an ever-increasing potential of these peptides. Undoubtedly, the most extensively studied bacteriocin is nisin, which has gained widespread applications in the food industry. This FDA-approved bacteriocin is produced by the GRAS microorganism Lactococcus lactis and is used as a food additive in at least 48 countries, particularly in processed cheese, dairy products and canned foods. Nisin is effective against food-borne pathogens such as L. monocytogenes and many other Gram-positive spoilage microorganisms. Nisin is listed in Spain as E-234, and may also be cited as nisin preservative or natural preservative. In addition to the work on nisin, several authors have outlined issues involved in the approval of new bacteriocins for food use.

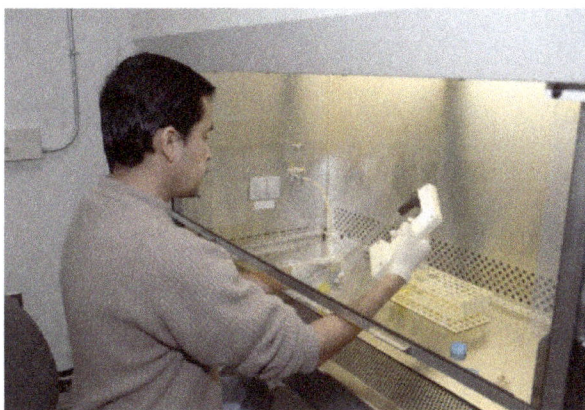

Hurdle Technology

"Hurdle Technology" is a technology that uses a combination of two or more preservation parameters at an optimum level in order to get a maximum lethality against micro-organisms without compromising with the nutritional and sensory qualities of food products. A wide range of preservation techniques is available e.g. freezing, blanching, pasteurizing and canning but the spoilage and poisoning of foods by microorganisms is a problem that is not yet under adequate control. Moreover, today's consumer demands for more natural and fresh-like foods, that requires the use of only mild preservation techniques. Hence, for the benefit of food manufacturers and fulfill the requirement of the consumer, there is a strong need for new or improved mild preservation methods that allow for the production of fresh-like, but stable and safe food. This topic would be primarily focussing on hurdle technology in food preservation.

Hurdle Technology in Food Preservation

Types of Hurdles

There are more than 60 potential hurdles those can be used for food preservation but the most important hurdles used are:

1. Temperature(high or low)

2. Water Activity (aw)

3. Acidity(ph)

4. Redox Potential (Eh)

5. Chemical Preservatives (Ex-nitrite, sorbate, sulfide)

6. Competitive Microorganisms (Ex-lactic acid bacteria)

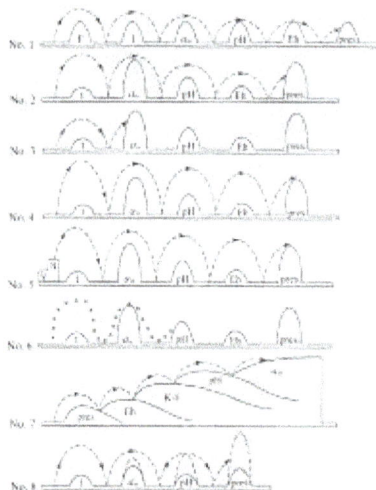

The preservation of almost all the foods is a result of combined application of several preservative methods (e.g., heating, chilling, drying, curing, conserving, acidification, oxygen-removal, fermenting, adding preservatives, etc.). These methods and their underlying principles (i.e., F, t, aw, pH, Eh, competitive flora, preservatives, etc.) have been applied since long back in our traditional foods empirically, but as the knowledge about these factors has increased, they are being applied intelligently using the concept of hurdle technology. Recently several novel preservative techniques (e.g. gas packaging, biopreservation, bacteriocins, ultra high-pressure treatment, edible coatings, etc.) have gained popularity to be used in combination with other traditional preservative factors(Hurdles).

Every hurdle could have both, positive or a negative effect on foods, depending on its intensity. For example, use of low temperature (chilling) below the critical limit of any food can lead to "chilling injury" whereas moderate chilling will be beneficial for their shelf life extension as it retards microbial growth. Similarly lowering the pH in fermented sausage inhibits the growth of pathogenic bacteria but lowering beyond the required limit can also impair the taste. Therefore a balanced intensity of any hurdle should be used for food preservation. The Too small intensity of hurdle should be strengthened and if it is detrimental to food quality, it should be lowered. This adjustment will provide a food that would be safe as well as will have a good quality by the use of hurdles in an optimal range.

Mechanism of Food Preservation by Hurdle Technology

The whole mechanism of preservation of food by using the concept of hurdle technology is comprised of various responses those are given by any microorganism. The whole phenomenon can be understood by following:

- Homeostasis: Tendency of any organism to maintain its internal status is known as Homeostasis. The homeostasis of microorganisms plays a key role in food preservation. If any of the hurdles used in food disturbs the homeostasis of these microorganisms, they will not be able to multiply and will remain constant in number or will die before the re-establishment of homeostasis. Therefore, food preservation can be achieved by disturbing the homeostasis of microorganism, temporarily or permanently;

- Metabolic Exhaustion: Another important phenomenon for food preservation is Metabolic Exhaustion of microorganisms. As a response to the hurdles applied to foods, microorganisms try to repair their homeostasis, use up all their energy for this and become metabolically exhausted. This leads to an auto-sterilization of such foods. The foods which are preserved with the concept of hurdle technology and are microbiologically stable, become safer during storage at ambient temperature. The microbes can respond better to the hurdles at ambient temperature than at refrigeration and become metabolically exhausted;

- Stress Reactions: As a response of various hurdles e.g. heat, pH, aw , ethanol, oxidative compounds, etc. as well starvation, a stress shock protein is generated by some bacteria. These stress proteins may hamper food preservation and could turn out to be problematic for the application of hurdle technology if only one hurdle has been applied. If different stresses are received by the microorganism at the same time, the activation of genes for the synthesis of stress shock proteins, which help organisms to cope with stress situation, would be difficult. Synthesis of many stress shock proteins due to simultaneous exposure to different stresses will be very energy-consuming and would lead to metabolic exhaustion of the microorganism;

- Multi-Target Preservation: A combined effect could be achieved by hitting various targets within the microbial cell (e.g., cell membrane, DNA, enzyme systems, pH, aw, Eh) by using different hurdles simultaneously. This disturbs the homeostasis of the microorganisms present in several respects. In this case, the replenishment of homeostasis and activation of stress shock proteins becomes more difficult. Therefore, simultaneous application of different hurdles in a particular food would lead to optimal microbial stability.

References

- Rickus, Alexis; Saunder, Bev; Mackey, Yvonne (2016-08-22). AQA GCSE Food Preparation and Nutrition. Hodder Education. ISBN 9781471863653

- Cooling-chilling, heating-cooling, processing-technology: hyfoma.com, Retrieved 30 May 2018

- Bhat R.; Sridhar K. R.; Tomita Y.; Tomita-Yokotanib K. (2007). "Effect of ionizing radiation on antinutritional features of velvet bean seeds (Mucuna pruriens)". Food Chemistry. 103: 860–866. doi:10.1016/j.foodchem.2006.09.037

- "Scientific Status Summary Irradiation of Food". Institute of Food Technologists' Expert Panel on Food Safety and Nutrition in Food Technology. January 1998. Retrieved May 30, 2015

- Salting-pickling-processes-food-preparation-preservation: foodsafetyhelpline.com, Retrieved 11 March 2018

- European Food Safety Authority (2011). "Statement summarising the Conclusions and Recommendations from the Opinions on the Safety of Irradiation of Food adopted by the BIOHAZ and CEF Panels". EFSA Journal. 9 (4): 2107

- Publishing, D. K. (2005-08-29). The Cook's Book: Techniques and tips from the world's master chefs. Penguin. ISBN 9780756665609

- Canning-food-processing: britannica.com, Retrieved 13 July 2018

- Trigo M. J.; Sousa M. B.; Sapata M. M.; Ferreira A.; Curado T.; Andrada L.; Veloso M. G. (2009). "Radiation processing of minimally processed vegetables and aromatic plants". Radiation Physics and Chemistry. 78: 659–663. doi:10.1016/j.radphyschem.2009.03.052

- "Food Irradiation in Asia, the European Union, and the United States" (PDF). Japan Radioisotope Association. May 2013. Archived from the original (PDF) on February 9, 2015. Retrieved January 6, 2015

- Food-preservation-methods-canning-freezing-and-drying, canning, food-drink: dummies.com, Retrieved 03 July 2018

- Fan Xuetong (2011). "Changes in Quality, Liking, and Purchase Intent of Irradiated Fresh-Cut Spinach during Storage". Journal of Food Science. 76: S363–S368. doi:10.1111/j.1750-3841.2011.02207.x

- Rennie, Richard (2016). boiling-point elevation. A Dictionary of Chemistry. Oxford: Oxford University Press. ISBN 9780198722823

- Hurdle-technology-in-food-preservation: discoverfoodtech.com, Retrieved 18 May 2018

Food Additives

A food additive is a substance that is used in food to preserve and enhance its appearance, flavor and taste. All the varied food additives and nutraceuticals as well as the processes of food fortification, food coloring and food coating have been covered in extensive detail in this chapter.

Food Fortification

Food fortification– also known as food enrichment– is when nutrients are added to food at higher levels than what the original food provides. This is done to address micronutrient deficiencies across populations, countries and regions.

Governments working with industry, international agencies and NGOs have used this method to help reduce and eliminate micronutrient deficiencies in their populations.

Fortification of centrally-processed staple foods is a simple, affordable and viable approach to reach large sections of a country's population with iron, folic acid, and other essential micronutrients.

Adding micronutrients to common staple foods can significantly improve the nutritional quality of the food supply and improve public health with minimal risk. The foods most commonly fortified are salt, wheat, corn, rice, bouillon cubes, soya sauce and other condiments.

Fortifying commonly-eaten grains such as wheat, maize flour and rice is among the easiest and least expensive ways to prevent disease, strengthen immune systems and nurture a healthy and productive next generation.

Currently, 79 countries around the world have made it the law to fortify at least one major grain: 78 of them fortify wheat flour, 12 fortify maize products and five fortify rice. These grains are usually fortified with vitamin A, iron and folic acid, which help prevent blindness, anaemia and birth defects, and improve mental function.

Nutrition International's Grain Fortification Programs

Nutrition International leads and supports grain fortification efforts in developing countries through a number of programs including:

- Partnering with the Global Alliance for Improved Nutrition (GAIN), UNICEF and the South African government to help large South African flour mills fortify maize and wheat flour with vitamin A, iron, and folic acid.

- Working with the World Food Programme (WFP) and flour millers in Pakistan to fortify flour for distribution in Afghanistan, where it reached about 2.5 million people.

- Supporting and expanding wheat and maize flour fortification programs, including national programs in Yemen, Iran, India, Pakistan, Nepal and Bolivia.

- Working with the Food Fortification Initiative, a network of public and private agencies, to increase flour fortification in developing countries.

- Advocating together with UNICEF, the U.S. Centers for Disease Control, and other organizations, to increase the number of countries routinely adding iron to flour from two in 1990 to 79 in 2014, including Central and South America and the Middle East, Indonesia, Nigeria and South Africa.

While less widespread than grain fortification, fortifying cooking oil with vitamin A and other micronutrients is also a simple and inexpensive way to fight vitamin A deficiency and disease. Cooking oil is an ideal carrier for micronutrients because it is so commonly used and the cost of fortification at the production stage is low.

Nutrition International supports the fortification of cooking oil in a number of ways:

- Training nutrition consultants from developing nations and giving them the tools to lobby their national governments and private industry to add vitamin A to cooking oils and grains.

- Working with cooking oil producers to show them how to easily fortify their products without significantly adding to the cost.

- Supporting governments in drafting legislation to make it mandatory to fortify cooking oil with vitamin A and wheat flour with iron and folic acid, leveling the playing field for all oil and grain producers.

- Helping governments, especially quality control authorities, in strengthening their capacity to put in place a strong quality control system and a viable enforcement mechanism.

- Helping cooking oil manufacturers upgrade their equipment for fortification and laboratories strengthening for internal quality control.

Food fortification can happen at the household level, the community level or, most commonly, at the industrial level:

- *Mass fortification* is when micronutrients are added to foods commonly consumed by the mass population – such as cereals and condiments.

- *Universal fortification* is when micronutrients are added to food consumed by animals as well as people, such as with iodization of salt.

- *Targeted fortification* exists in such areas as school food programs, when, for example, a cracker is specifically fortified for a targeted age group.

Fortification of foods is either mandatory, which means it is legislated by the government, or voluntary. Fortification of centrally-processed staples is a simple, affordable and viable approach to reaching major sections of the population with essential vitamins and minerals.

Examples of Fortification in Foods

Many foods and beverages worldwide have been fortified, whether a voluntary action by the product developers or by law. Although some may view these additions as strategic marketing schemes to sell their product, there is a lot of work that must go into a product before simply fortifying it. In order to fortify a product, it must first be proven that the addition of this vitamin or mineral is beneficial to health, safe, and an effective method of delivery. The addition must also abide by all food and labeling regulations and support nutritional rationale. From a food developer's point of view, they also need to consider the costs associated with this new product and whether or not there will be a market to support the change.

Examples of foods and beverages that have been fortified and shown to have positive health effects:

Iodized Salt

"Iodine deficiency disorder (IDD) is the single greatest cause of preventable mental retardation. Severe deficiencies cause cretinism, stillbirth and miscarriage. But even mild deficiency can significantly affect the learning ability of populations. Today over 1 billion people in the world suffer from iodine deficiency, and 38 million babies born every year are not protected from brain damage due to IDD."—Kul Gautam, Deputy Executive Director, UNICEF, October 2007.

Iodised salt has been used in the United States since before World War II. It was discovered in 1821 that goiters could be treated by the use of iodized salts. However, it was not until 1916 that the use of iodized salts could be tested in a research trial as a preventative measure against goiters. By 1924, it became readily available in the US. Currently in Canada and the US, the RDA for iodine is as low as 90 µg/day for children (4–8 years) and as high as 290 µg/day for breast-feeding mothers.

Diseases that are associated with an iodine deficiency include: mental retardation, hypothyroidism, and goiter. There is also a risk of various other growth and developmental abnormalities.

Folic Acid

Folic acid (also known as folate) functions in reducing blood homocysteine levels, forming red blood cells, proper growth and division of cells, and preventing neural tube defects (NTDs). In many industrialized countries, the addition of folic acid to flour has prevented a significant number of

NTDs in infants. Two common types of NTDs, spina bifida and anencephaly, affect approximately 2500-3000 infants born in the US annually. Research trials have shown the ability to reduce the incidence of NTDs by supplementing pregnant mothers with folic acid by 72%.

The RDA for folic acid ranges from as low as 150 μg/day for children aged 1–3 years old, to 400 μg/day for males and females over the age of 19, and 600 μg/day during pregnancy. Diseases associated with folic acid deficiency include: megaloblastic or macrocytic anemia, cardiovascular disease, certain types of cancer, and NTDs in infants.

Niacin

Niacin has been added to bread in the USA since 1938 (when voluntary addition started), a programme which substantially reduced the incidence of pellagra. As early as 1755, pellagra was recognized by doctors as being a niacin deficiency disease. Although not officially receiving its name of pellagra until 1771. Pellagra was seen amongst poor families who used corn as their main dietary staple. Although corn itself does contain niacin, it is not a bioavailable form unless it undergoes nixtamalization (treatment with alkali, traditional in Native American cultures) and therefore was not contributing to the overall intake of niacin.

The RDA for niacin is 2 mg NE(niacin equivalents)/day (AI) for infants aged 0–6 months, 16 mg NE/day for males, and 14 mg NE/day for females who are over the age of 19.

Diseases associated with niacin deficiency include: Pellagra which consisted of signs and symptoms called the 3D's-"Dermatitis, dementia, and diarrhea. Others may include vascular or gastrointestinal diseases. Common diseases which present a high frequency of niacin deficiency: alcoholism, anorexia nervosa, HIV infection, gastrectomy, malabsorptive disorders, certain cancers and their associated treatments.

Vitamin D

Since Vitamin D is a fat-soluble vitamin, it cannot be added to a wide variety of foods. Foods that it is commonly added to are margarine, vegetable oils and dairy products. During the late 1800s, after the discovery of curing conditions of scurvy and beriberi had occurred, researchers were aiming to see if the disease, later known as rickets, could also be cured by food. Their results showed that sunlight exposure and cod liver oil were the cure. It was not until the 1930s that vitamin D was actually linked to curing rickets. This discovery led to the fortification of common foods such as milk, margarine, and breakfast cereals. This took the astonishing statistics of approximately 80–90% of children showing varying degrees of bone deformations due to vitamin D deficiency to being a very rare condition.

The current RDA for infants aged 0–6 months is 10 μg (400 International Units (IU))/day and for adults over 19 years of age it is 15 μg (600 IU)/day.

Diseases associated with a vitamin D deficiency include rickets, osteoporosis, and certain types of cancer (breast, prostate, colon and ovaries). It has also been associated with increased risks for fractures, heart disease, type 2 diabetes, autoimmune and infectious diseases, asthma and other wheezing disorders, myocardial infarction, hypertension, congestive heart failure, and peripheral vascular disease.

Fluoride

Although fluoride is not considered an essential mineral, it is useful in prevention of tooth decay and maintaining adequate dental health. In the mid-1900s it was discovered that towns with a high level of fluoride in their water supply was causing the residents' teeth to have both brown spotting and a strange resistance to dental caries. This led to the fortification of water supplies with fluoride in safe amounts (or reduction of naturally-occurring levels) to retain the properties of resistance to dental caries but avoid the staining cause by fluorosis (a condition caused by excessive fluoride intake).The tolerable upper intake level (UL) set for fluoride ranges from 0.7 mg/day for infants aged 0–6 months and 10 mg/day for adults over the age of 19.

Conditions commonly associated with fluoride deficiency are dental caries and osteoporosis.

Others

Some other examples of fortified foods:

- Calcium is frequently added to fruit juices, carbonated beverages and rice.

- White rice is frequently enriched to replace some of the lost nutrients during milling or adding extras in.

- Amylase rich flour is utilized for food making to increase dietary consumption.

- Zinc is found to improve its blood level when used alone for fortification but more studies are needed to assess other benefits.

Biofortification

Biofortification is the process by which the nutritional quality of food crops is improved through agronomic practices, conventional plant breeding, or modern biotechnology. Biofortification differs from conventional fortification in that biofortification aims to increase nutrient levels in crops during plant growth rather than through manual means during processing of the crops. Biofortification may therefore present a way to reach populations where supplementation and conventional fortification activities may be difficult to implement and/or limited.

Food Additives

Food additives are any of various chemical substances added to foods to produce specific desirable effects. Additives such as salt, spices, and sulfites have been used since ancient times to preserve foods and make them more palatable. With the increased processing of foods in the 20th century, there came a need for both the greater use of and new types of food additives. Many modern products, such as low-calorie, snack, and ready-to-eat convenience foods, would not be possible without food additives.

MSGLearn about the myths and safety of monosodium glutamate (MSG).

There are four general categories of food additives: nutritional additives, processing agents, preservatives, and sensory agents. These are not strict classifications, as many additives fall into more than one category.

Nutritional Additives

Nutritional additives are used for the purpose of restoring nutrients lost or degraded during production, fortifying or enriching certain foods in order to correct dietary deficiencies, or adding nutrients to food substitutes. The fortification of foods began in 1924 when iodine was added to table salt for the prevention of goitre. Vitamins are commonly added to many foods in order to enrich their nutritional value. For example, vitamins A and D are added to dairy and cereal products, several of the B vitamins are added to flour, cereals, baked goods, and pasta, and vitamin C is added to fruit beverages, cereals, dairy products, and confectioneries. Other nutritional additives include the essential fatty acid linoleic acid, minerals such as calcium and iron, and dietary fibre.

Processing Agents

A number of agents are added to foods in order to aid in processing or to maintain the desired consistency of the product.

Processing additives and their uses		
Function	**Typical Chemical Agent**	**Typical Product**
anticaking	sodium aluminosilicate	salt
bleaching	benzoyl peroxide	flour

Processing additives and their uses		
chelating	ethylenediaminetetraacetic acid (EDTA)	dressings, mayonnaise, sauces, dried bananas
clarifying	bentonite, proteins	fruit juices, wines
conditioning	potassium bromate	flour
emulsifying	lecithin	ice cream, mayonnaise, bakery products
leavening	yeast, baking powder, baking soda	bakery products
moisture control (humectants)	glycerol	marshmallows, soft candies, chewing gum
pH control	citric acid, lactic acid	certain cheeses, confections, jams and jellies
stabilizing and thickening	pectin, gelatin, carrageenan, gums (arabic, guar, locust bean)	dressings, frozen desserts, confections, pudding mixes, jams and jellies

Emulsifiers are used to maintain a uniform dispersion of one liquid in another, such as oil in water. The basic structure of an emulsifying agent includes a hydrophobic portion, usually a long-chain fatty acid, and a hydrophilic portion that may be either charged or uncharged. The hydrophobic portion of the emulsifier dissolves in the oil phase, and the hydrophilic portion dissolves in the aqueous phase, forming a dispersion of small oil droplets. Emulsifiers thus form and stabilize oil-in-water emulsions (e.g., mayonnaise), uniformly disperse oil-soluble flavour compounds throughout a product, prevent large ice crystal formation in frozen products (e.g., ice cream), and improve the volume, uniformity, and fineness of baked products.

Stabilizers and thickeners have many functions in foods. Most stabilizing and thickening agents are polysaccharides, such as starches or gums, or proteins, such as gelatin. The primary function of these compounds is to act as thickening or gelling agents that increase the viscosity of the final product. These agents stabilize emulsions, either by adsorbing to the outer surface of oil droplets or by increasing the viscosity of the water phase. Thus, they prevent the coalescence of the oil droplets, promoting the separation of the oil phase from the aqueous phase (i.e., creaming). The formation and stabilization of foam in a food product occurs by a similar mechanism, except that the oil phase is replaced by a gas phase. The compounds also act to inhibit the formation of ice or sugar crystals in foods and can be used to encapsulate flavour compounds.

Chelating, or sequestering, agents protect food products from many enzymatic reactions that promote deterioration during processing and storage. These agents bind to many of the minerals that are present in food (e.g., calcium and magnesium) and are required as cofactors for the activity of certain enzymes.

Preservatives

Food preservatives are classified into two main groups: antioxidants and antimicrobials. Antioxidants are compounds that delay or prevent the deterioration of foods by oxidative mechanisms. Antimicrobial agents inhibit the growth of spoilage and pathogenic microorganisms in food.

Food preservatives	
chemical agent	mechanism of action
Antioxidants	
ascorbic acid	oxygen scavenger
butylated hydroxyanisole (BHA)	free radical scavenger
butylated hydroxytoluene (BHT)	free radical scavenger
citric acid	enzyme inhibitor/metal chelator
sulfites	enzyme inhibitor/oxygen scavenger
tertiary butylhydroquinone (TBHQ)	free radical scavenger
tocopherols	free radical scavenger
Antimicrobials	
acetic acid	disrupts cell membrane function (bacteria, yeasts, some molds)
benzoic acid	disrupts cell membrane function/inhibits enzymes (molds, yeasts, some bacteria)
natamycin	binds sterol groups in fungal cell membrane (molds, yeasts)
nisin	disrupts cell membrane function (gram-positive bacteria, lactic acid-producing bacteria)
nitrates, nitrites	inhibits enzymes/disrupts cell membrane function (bacteria, primarily Clostridiumbotulinum)
propionic acid	disrupts cell membrane function (molds, some bacteria)
sorbic acid	disrupts cell membrane function/inhibits enzymes/inhibits bacterial spore germination (yeasts, molds, some bacteria)
sulfites and sulfur dioxide	inhibits enzymes/forms addition compounds (bacteria, yeasts, molds)

Antioxidants

The oxidation of food products involves the addition of an oxygen atom to or the removal of a hydrogen atom from the different chemical molecules found in food. Two principal types of oxidation that contribute to food deterioration are autoxidation of unsaturated fatty acids (i.e., those containing one or more double bonds between the carbon atoms of the hydrocarbon chain) and enzyme-catalyzed oxidation.

The autoxidation of unsaturated fatty acids involves a reaction between the carbon-carbon double bonds and molecular oxygen (O_2). The products of autoxidation, called free radicals, are highly reactive, producing compounds that cause the off-flavours and off-odours characteristic of oxidative rancidity. Antioxidants that react with the free radicals (called free radical scavengers) can slow the rate of autoxidation. These antioxidants include the naturally occurring tocopherols (vitamin E derivatives) and the synthetic compounds butylated hydroxyanisole (BHA), butylated hydroxytoluene (BHT), and tertiary butylhydroquinone (TBHQ).

Specific enzymes may also carry out the oxidation of many food molecules. The products of these oxidation reactions may lead to quality changes in the food. For example, enzymes called phenolases catalyze the oxidation of certain molecules (e.g., the amino acid tyrosine) when fruits and vegetables, such as apples, bananas, and potatoes, are cut or bruised. The product of these oxidation reactions, collectively known as enzymatic browning, is a dark pigment called melanin. Antioxidants that inhibit enzyme-catalyzed oxidation include agents that bind free oxygen (i.e., reducing agents), such as ascorbic acid (vitamin C), and agents that inactivate the enzymes, such as citric acid and sulfites.

Antimicrobials

Antimicrobials are most often used with other preservation techniques, such as refrigeration, in order to inhibit the growth of spoilage and pathogenic microorganisms. Sodium chloride (NaCl), or common salt, is probably the oldest known antimicrobial agent. Organic acids, including acetic, benzoic, propionic, and sorbic acids, are used against microorganisms in products with a low pH. Nitrates and nitrites are used to inhibit the bacterium *Clostridium botulinum* in cured meat products (e.g., ham and bacon). Sulfur dioxide and sulfites are used to control the growth of spoilage microorganisms in dried fruits, fruit juices, and wines. Nisin and natamycin are preservatives produced by microorganisms. Nisin inhibits the growth of some bacteria, while natamycin is active against molds and yeasts.

Sensory Agents

Colorants

Colour is an extremely important sensory characteristic of foods; it directly influences the perception of both the flavour and quality of a product. The processing of food can cause degradation **or** loss of natural pigments in the raw materials. In addition, some formulated products, such as soft drinks, confections, ice cream, and snack foods, require the addition of colouring agents. Colorants are often necessary to produce a uniform product from raw materials that vary in colour intensity. Colorants used as food additives are classified as natural or synthetic. Natural colorants are derived from plant, animal, and mineral sources, while synthetic colorants are primarily petroleum-based chemical compounds.

soft drink many soft drinks, including colas,

Natural Colorants

Most natural colorants are extracts derived from plant tissues. The use of these extracts in the food industry has certain problems associated with it, including the lack of consistent colour intensities, instability upon exposure to light and heat, variability of supply, reactivity with other food components, and addition of secondary flavours and odours. In addition, many are insoluble in water and therefore must be added with an emulsifier in order to achieve an even distribution throughout the food product.

Natural food colorants				
*Plus other similar compounds. **Many carotenoids used as food colorants are chemically synthesized.				
chemical class	colour	plant source	pigment	products
anthocyanins	red	strawberry (Fragaria species)	pelargonidin 3-glucoside*	beverages, confections, preserves, fruit products
	blue	grape (Vitis species)	malvidin 3-glucoside*	beverages
betacyanins	red	beetroot (Beta vulgaris)	betanin	dairy products, desserts, icings
carotenoids**	yellow/orange	annatto (Bixa orellana)	bixin	dairy products, margarine
	yellow	saffron (Crocus sativus)	crocin	rice dishes, bakery products
	red/orange	paprika (Capsicum annuum)	capsanthin	soups, sauces
	orange	carrot (Daucus carota)	beta-carotene	bakery products, confections
	red	mushroom (Cantharellus cinnabarinus)	canthaxanthin	sauces, soups, dressings
phenolics	orange/yellow	turmeric (Cuycuma longa)	curcumin	dairy products, confections

Synthetic Colorants

Synthetic colorants are water-soluble and are available commercially as powders, pastes, granules, or solutions. Special preparations called lakes are formulated by treating the colorants with aluminum hydroxide. They contain approximately 10 to 40 percent of the synthetic dye and are insoluble in water and organic solvents. Lakes are ideal for use in dry and oil-based products. The stability of synthetic colorants is affected by light, heat, pH, and reducing agents. A number of dyes have been chemically synthesized and approved for usage in various countries. These colorants are designated according to special numbering systems specific to individual countries. For example, the United States uses FD&C numbers (chemicals approved for use in foods, drugs, and cosmetics), and the European Union (EU) uses E numbers.

Synthetic food colorants			
common name	designation		products
	United States	European Union	
allura red AC	FD&C red no. 40	. . .	gelatin, puddings, dairy products, confections, beverages
brilliant blue FCF	FD&C blue no. 1	E133	beverages, confections, icings, syrups, dairy products
erythrosine	FD&C red no. 3	E127	maraschino cherries
fast green FCF	FD&C green no. 3	. . .	beverages, puddings, ice cream, sherbet, confections
indigo carmine	FD&C blue no. 2	E132	confections, ice cream, bakery products
sunset yellow FCF	FD&C yellow no. 6	E110	bakery products, ice cream, sauces, cereals, beverages
tartrazine	FD&C yellow no. 5	E102	beverages, cereals, bakery products, ice cream, sauces

All synthetic colorants have undergone extensive toxicological analysis. Brilliant Blue FCF, Indigo Carmine, Fast Green FCF, and Erythrosine are poorly absorbed and show little toxicity. Extremely high concentrations (greater than 10 percent) of Allura Red AC cause psychotoxicity, and Tartrazine induces hypersensitive reactions in some persons. Synthetic colorants are not universally approved in all countries.

Flavourings

The flavour of food results from the stimulation of the chemical senses of taste and smell by specific food molecules. Taste reception is carried out in specialized cells located in the taste buds. The five basic taste sensations—sweet, salty, bitter, sour, and umami—are detected in regions of the tongue, mouth, and throat. Taste cells are specific for certain flavour molecules (e.g., sweeteners).

In addition to the basic tastes, the flavouring molecules in food stimulate specific olfactory (smell) cells in the nasal cavity. These cells can detect more than 10,000 different stimuli, thus fine-tuning the flavour sensation of a food.

A flavour additive is a single chemical or blend of chemicals of natural or synthetic origin that provides all or part of the flavour impact of a particular food. These chemicals are added in order to replace flavour lost in processing and to develop new products. Flavourings are the largest group of food additives, with more than 1,200 compounds available for commercial use. Natural flavourings are derived or extracted from plants, spices, herbs, animals, or microbial fermentations. Artificial flavourings are mixtures of synthetic compounds that may be chemically identical to natural flavourings. Artificial flavourings are often used in food products because of the high cost, lack of availability, or insufficient potency of natural flavourings.

Flavour enhancers are compounds that are added to a food in order to supplement or enhance its own natural flavour. The concept of flavour enhancement originated in Asia, where cooks added seaweed to soup stocks in order to provide a richer flavour to certain foods. The flavour-enhancing component of seaweed was identified as the amino acid L-glutamate, and monosodium glutamate(MSG) became the first flavour enhancer to be used commercially. The rich flavour associated with L-glutamate was called umami.

Other compounds that are used as flavour enhancers include the 5'-ribonucleotides, inosine monophosphate (IMP), guanosine monophosphate (GMP), yeast extract, and hydrolyzed vegetableprotein. Flavour enhancers may be used in soups, broths, sauces, gravies, flavouring and spice blends, canned and frozen vegetables, and meats.

Sweeteners

Sucrose, or table sugar, is the standard on which the relative sweetness of all other sweeteners is based. Because sucrose provides energy in the form of carbohydrates, it is considered a nutritive sweetener. Other nutritive sweeteners include glucose, fructose, corn syrup, high-fructose corn syrup, and sugar alcohols (e.g., sorbitol, mannitol, and xylitol).

sugar; corn syrup

Efforts to chemically synthesize sweeteners began in the late 1800s with the discovery of saccharin. Since then, a number of synthetic compounds have been developed that provide few or no calories or nutrients in the diet and are called nonnutritive sweeteners. These sweeteners have significantly greater sweetening power than sucrose, and therefore a relatively low concentration may be used in food products. In addition to saccharin, the most commonly used nonnutritive sweeteners are cyclamates, aspartame, and acesulfame K.

aspartame discover the science behind the safety of aspartame.

The sensation of sweetness is transmitted through specific protein molecules, called receptors, located on the surface of specialized taste cells. All sweeteners function by binding to these receptors on the outside of the cells. The increased sweetness of the nonnutritive sweeteners relative to sucrose may be due to either tighter or longer binding of these synthetic compounds to the receptors.

Nonnutritive sweeteners are primarily used for the production of low-calorie products including baked goods, confectioneries, dairy products, desserts, preserves, soft drinks, and tabletop sweeteners. They are also used as a carbohydrate replacement for persons with diabetes mellitus and in chewing gum and candies to minimize the risk of dental caries (i.e., tooth decay). Unlike nutritive sweeteners, nonnutritive sweeteners do not provide viscosity or texture to products, so bulking agents such as polydextrose are often required for manufacture.

Toxicological Testing and Health Concerns

Food additives and their metabolites are subjected to rigorous toxicological analysis prior to their approval for use in the industry. Feeding studies are carried out using animal species (e.g., rats, mice, dogs) in order to determine the possible acute, short-term, and long-term toxic effects of these chemicals. These studies monitor the effects of the compounds on the behaviour, growth, mortality, blood chemistry, organs, reproduction, offspring, and tumour development in the test animals over a 90-day to two-year period. The lowest level of additive producing no toxicological effects is called the no-effect level (NOEL). The NOEL is generally divided by 100 to determine a maximum acceptable daily intake (ADI).

Toxicological analysis of the nonnutritive sweeteners has produced variable results. High concentrations of saccharin and cyclamates in the diets of rats have been shown to induce the development of bladder tumours in the animals. Because of these results, the use of cyclamates has been banned in several countries, including the United States, and the use of saccharin must include a qualifying statement regarding its potential health risks. However, no evidence of human bladder cancer has been reported with the consumption of these sweeteners. Both aspartame and acesulfame K have been deemed to be relatively safe, with no evidence of carcinogenic potential in animal studies.

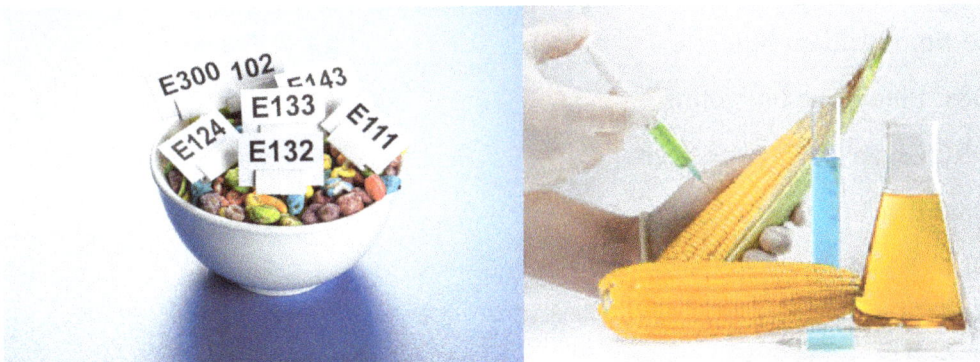

Food Coloring

It is added to food and drinks to create a specific appearance.

Food colors are prevalent in our daily life and are even in foods and drinks we wouldn't expect. They're used to make the orange color of oranges brighter and more uniform, to create the brown hues in colas, and blue dye is added to marshmallows to make them, strangely, whiter! Of course, there are the obvious uses - to decorate a cake or cupcakes, candy (think M&Ms and Skittles) and the countless multi-colored foods and drinks we're so accustomed to seeing. They're in so many foods, in fact, we hardly think twice about how they became so vivid and colorful in the first place, and most of us are still unaware of the potential dangers associated with them.

Food coloring is also known as artificial color.

Artificial colors contain various chemicals and are commonly derived from petroleum products. Conveniently, they are available in various forms, including liquids, powders, gels, and pastes.

Read more about artificial colors where you may not expect them

In the U.S., all food and drink labels must list the artificial colors they contain. To denote synthetic food coloring agents (or artificial dyes), they are assigned FD&C (Federal Food, Drug and Cosmetic) numbers which are regulated by the FDA.

In the U.S., these seven artificial colors are approved for use in food:

FD&C Blue No. 1 Brilliant Blue

- FD&C Blue No. 2 Indigotine
- FD&C Green No. 3 Fast Green
- FD&C Red No. 40 Allura Red
- FD&C Red No. 3 Erythrosine
- FD&C Yellow No. 5 Tartrazine
- FD&C Yellow No. 6 Sunset Yellow

These are known as primary colors. When they are mixed to produce other colors, those colors are known as secondary colors.

Two additional colors, Orange B and Citrus Red 2 are allowed for specific foods only. Orange B is used only for hot dog and sausage casings. Citrus Red 2 is used only to color orange peels.

Safety Concerns

Although artificial colors have been linked to allergic reactions and other health concerns, including behavioral changes in children, the FDA continues to allow them to be used in food. One of the most problematic dyes is Yellow #5, E102, tartrazine, used to color soft drinks and energy drinks, cake mixes, salty snacks, cereals, packaged soups and more. The coal tar dye has been linked to cancer and is known to provoke asthma attacks, skin reactions, and hyperactivity in children. Tartrazine has already been banned in Norway, Austria and Finland.

Natural Food Coloring

Natural food colors are exempt from certification. They are derived from natural sources, for example:

- Annatto - a reddish-orange dye made from a South American shrub
- Paprika
- Saffron
- Lycopene from tomatoes
- Beta-carotene
- Carmine, derived from the cochineal insect.

Though they are known as natural dyes, hexane, acetone, and other solvents are used to break down the cell walls in the fruit or vegetable they are derived from to extract the colors. Another issue is that some natural food dyes are linked with serious allergic reactions, as well as other health concerns.

Artificial Colors for Food Coloring

Since labels indicate which artificial colors artificial and natural- are included in the ingredients, carefully read the ingredients list on every product you buy. Remember, food dyes are hiding where you may not expect them. If a discernible reaction occurs with a food or drink containing a food dye, record the color or combination of colors and watch for that reaction again. While most people react to one specific color (Red No. 40 is a culprit well known amongst parents) some people react to a combination of dyes.

Naturally occurring color additives from vegetable and mineral sources were used to color foods, drugs, and cosmetics in ancient times. Paprika, turmeric, saffron, iron and lead oxides, and copper sulfate are some examples. The early Egyptians used artificial colors in cosmetics and hair dyes. Wine was artificially colored beginning in at least 300 BC.

In 1856, William Henry Perkin discovered the first synthetic organic dye, called mauve. Discoveries of similar dyes soon followed and they quickly became used to color foods, drugs, and cosmetics. Because these dyes were first produced from by-products of coal processing, they were known as "coal-tar colors."

Federal oversight of color additives began in the1880s. The assessment of color-imparting ingredients in foods was among the first public initiatives undertaken by the U.S. when, in 1881, the U.S. Department of Agriculture's (USDA) Bureau of Chemistry began research on the use of colors in food. Butter and cheese were the first foods for which the federal government authorized the use of artificial coloring.

By 1900, many foods, drugs, and cosmetics available in the U.S. were artificially colored. However, not all of the coloring agents were harmless and some were being used to hide inferior or defective foods. A careful assessment of the chemicals used for coloring foods at the time found many blatantly poisonous materials such as lead, arsenic, and mercury being added. In many cases, the toxicities of the starting materials for synthesizing coloring agents were well known and could be toxins, irritants, sensitizers, or carcinogens.

In 1906, Congress passed the Food and Drugs Act, which prohibited the use of poisonous or deleterious colors in confectionery and the coloring or staining of food to conceal damage or inferiority. The USDA had initial enforcement authority for this In 1907, the USDA issued Food Inspection Decision (F.I.D.) 76, which contained a list of seven straight colors approved for use in food. Subsequent F.I.D.'s in the early part of the century established a voluntary certification program and listed new colors.

In 1927, responsibility for enforcing the Food and Drugs Act of 1906 was given to the newly established FDA. (The agency was first called the Food, Drug, and Insecticide Administration and was given its current name in 1930.) By 1931, there were 15 straight colors approved for use in food, including six of the seven in use today: FD&C Blue No. 1 (Brilliant Blue FCF), FD&C Blue No. 2 (Indigotine), FD&C Green No. 3 (Fast Green FCF), FD&C Red No. 3 (Erythrosine), FD&C Yellow No. 5 (Tartrazine), and FD&C Yellow No. 6 (Sunset Yellow).

Federal Food, Drug, and Cosmetic Act of 1938. In the 1920s and 1930s, it became clear that the Food and Drugs Act of 1906 did not go far enough to protect the public health from misbranded, adulterated, and even toxic products, including an eyelash dye that blinded some women. The Federal Food, Drug, and Cosmetic Act of 1938 further increased government oversight of food and drugs and, for the first time, passed legislation for the regulation of cosmetics and medical devices.

For color additives, the 1938 FD&C Act mandated the listing of those coal-tar colors (other than coal-tar hair dyes) that were "harmless and suitable" for use in foods, drugs, and cosmetics. In addition, the act: contained adulteration and misbranding provisions for the use of coal-tar colors in foods, drugs, and cosmetics; required the listing of new colors; and made mandatory the previously voluntary certification program for batches of listed colors, with associated fees. Color additive lakes were in use by this time and were included in the provisions of the 1938 FD&C Act. The initial listing of lakes for food use under the act restricted their use to coloring shell eggs (egg dyeing).

In response to the 1938 Act, through public hearings FDA created the FD&C, D&C, and Ext. D&C nomenclature for certifiable color additives. FDA also established labeling and recordkeeping provisions, identified diluents that could be added to color additives, and established procedures for requesting certification of color additives and adding new color additives to the permitted list.

Color Additive Amendments of 1960. In the fall of 1950, many children became ill from eating an orange Halloween candy containing 1-2% FD&C Orange No. 1, a color additive approved for use in food. That same year, U.S. House Representative James Delaney began holding hearings on the possible carcinogenicity of pesticide residues and food additives. These events prompted FDA to reevaluate all of the listed color additives. In the next few years, FDA found that several caused serious adverse effects and proceeded to terminate their listings. During that time, it also became clear that coal was no longer the primary raw material source for the manufacture of color additives.

The Color Additive Amendments of 1960 defined "color additive" and required that only color additives (except coal-tar hair dyes) listed as "suitable and safe" for a given use could be used in foods, drugs, cosmetics, and medical devices. The 1960 Amendments prescribed the factors that FDA must consider in determining whether a proposed use of a color additive is safe, as well as the specific conditions for safe use that must be included in the listing regulation. FDA updated the procedural regulations for the petition process in response to these amendments. Under these amendments, the approximately 200 color additives that were in commercial use at the time were provisionally listed and could be used on an interim basis until they were either permanently listed or terminated due to safety concerns or lack of commercial interest. Permanently listing a color additive for a proposed use was prohibited unless scientific data established its safety.

The 1960 Amendments also contained a "Delaney Clause" that prohibited the listing of a color additive shown to be a carcinogen. The clause states that "A color additive shall be deemed unsafe. if the additive is found. to induce cancer when ingested by man or animal, or after other relevant exposure of man or animal to such additive."

After 1960, FDA gradually removed color additives from the provisional list either by permanent listing or by termination of listing. Today about half of the "1960" color additives remain listed; only color additive lakes remain provisionally listed and initiatives are underway to permanently list them.

Regulation of Color Additives

FDA has regulatory oversight for color additives used in foods, drugs, cosmetics, and medical devices. FDA lists new color additives or new uses for listed color additives that have been shown to be safe for their intended uses in the Code of Federal Regulations (CFR), conducts a certification program for batches of color additives that are required to be certified before sale, and monitors the use of color additives in products in the U.S., including product labeling. These activities stem from FDA's role in enforcing the color additive provisions of the FD&C Act, the Fair Packaging and Labeling Act, and other applicable laws, including the recently enacted Public Health Security and Bioterrorism Preparedness and Response Act of 2002 that requires domestic and foreign manufacturers of color additives used as ingredients in foods to register with FDA by December 12, 2003.

Color additives used in foods, drugs, cosmetics, and medical devices must comply with individual listing regulations issued by FDA. The use of an unlisted color additive, the improper use of a listed color additive, or the use of a color additive that does not conform to the purity and identity specifications of the listing regulation may cause a product to be adulterated according to the provisions of the FD&C Act. FDA may take enforcement action against such products. Most products contain only a small amount of color additive, so it takes only a small quantity to potentially adulterate a large amount of product.

FDA has established regulations for color additives in Title 21 of the CFR, parts 70-82. The regulations in 21 CFR parts 73, 74, and 82 identify each listed color additive, provide chemical specifications for the color additives, and identify uses and restrictions, labeling requirements and the requirement for certification. The regulations in 21 CFR part 71 describe the premarket approval process for new color additives or new uses for listed color additives. 21 CFR part 80 pertains to color additive certification.

Additional regulations that provide specific requirements for color additives in foods, drugs, cosmetics, and medical devices are found in other parts of the CFR. For example, the labeling of food products is found at 21 CFR 101.22(k) and cosmetic products at 21 CFR 701.3. Color additives are sometimes called "artificial color" or "artificial coloring" (21 CFR 101.22(a)(4)). From the regulatory standpoint, the term "colorant" refers to a dye or pigment used in a food contact material such as a polymer and doesn't migrate to food. These materials are regulated not as color additives but as food additives (21 CFR 178.3297(a)). Much of FDA's basic information on color additives is available online.

Listed Color Additives

All color additives required to be listed by FDA fall into two categories: those that are subject to FDA's certification process and those that are exempt from the certification process. Color additives subject to batch certification are synthetic organic dyes, lakes, or pigments. Those for food use are chemically classified as azo, xanthene, triphenylmethane, and indigoid dyes. Although certifiable color additives have been called coal-tar colors because of their traditional origins, today they are synthesized mainly from raw materials obtained from petroleum.

Color additives exempt from certification generally include those derived from plant or mineral sources. One, cochineal extract (and its lake, carmine) is derived from an insect. Most are straight colors; one exception is carmine as described above. Certification exempt color additives must comply with the identity and purity specifications and use limitations described in their listing regulations. Users of these color additives are responsible for ensuring that the color additives comply with the listing regulations.

Straight colors subject to batch certification are listed in 21 CFR part 74 and takes subject to batch certification as listed in 21 CFR part 82. Color additives exempt from certification are listed in 21 CFR part 73. Table gives the complete list of straight colors permitted for use in foods. More information on listed color additives is given on FDA's Web site.

Color Additives and GRAS

The FD&C Act provides for an exemption of some substances from the definition of "food additive" if they are generally recognized as safe for their intended uses. Such an exemption does not apply to color additives. However, a substance that is listed as GRAS also may be listed as a color additive.

A mixture of carotenoid xanthophyll esters ("lutein esters") is the subject of a recent GRAS notice submitted to FDA in support of its use as a food ingredient. The compound is dark orange-brown and may be capable of imparting color to a food. FDA's response letter to the notice reminds the manufacturer that use of the substance as a color additive, in addition to use as a GRAS substance, would require premarket approval by FDA.

Color Additive Certification

Color additive certification is the process by which FDA assures that newly manufactured batches of color additives meet the identity and specification requirements of their listing regulations. During fiscal year 2002, FDA certified batches representing a total of 16.5 million pounds of color additives, much of it for food uses.

The decision about the need for batch certification is made during the agency's review of a petition requesting a listing for the color additive. Batch certification is required when the composition needs to be controlled to protect the public health. Some color additives may contain impurities of toxicological concern, such as carcinogenic constituents.

The requirements for color additive certification, as well as storage, fees, recordkeeping, and inspection for owners and manufacturers, are described in detail in 21 CFR part 80. Regulations in 21 CFR part 70.25 prescribe labeling requirements for color additive batches before and after certification. Under the certification process, a sample from each manufactured batch of a certifiable color additive must be sent to FDA's Color Certification Branch accompanied by a "Request for Certification" that provides information about the batch including the name of the color additive, the name of the manufacturer, the batch weight, storage conditions for the batch, and the use for which it is being certified. FDA charges a fee for certification based on the batch weight. Prior to certification, the batch cannot be used in food, drug, cosmetic, or medical device products and must be stored separately from batches already certified.

Upon receipt of the sample, FDA personnel evaluate its physical appearance and chemically analyze it. At least 10 analyses are performed, for purity (total color content), moisture, residual salts, unreacted intermediates, colored impurities other than the main color (called subsidiary colors), any other specified impurities, and the heavy metals lead, arsenic, and mercury. The evaluation and analyses typically take less than five working days. The results are reviewed for compliance with the identity and specifications described in the listing regulation for the color additive. If the sample is found to meet these requirements, FDA issues a certificate for the batch that identifies the color additive, the batch weight, the uses for which the color additive is certified, the name and address of the owner, and other information as required. FDA also assigns a unique lot number for the batch and the name of the batch changes. For example, a batch of "tartrazine" becomes "FD&C Yellow No. 5."

Analytical and informational components of the certification program have been automated to the fullest extent possible. Currently, an on-line web-based system allows color additive manufacturers to submit and access information about individual samples, including receipt of FDA's certificates. Owners of certified batches are subject to FDA inspections of their establishments. During these inspections, FDA examines records of use of the color additives and takes samples from certified batches for analysis for comparison with FDA's original results.

Petition Review Process

When evaluating the safety of a new color additive or a new use for a listed color additive, FDA considers such factors as probable consumption or exposure from its use, cumulative effect in the diet, evaluation by experts qualified by scientific training and experience, and the availability of analytical methods for determining its purity and acceptable levels of impurities.

Any interested person may petition FDA for the use of a new color additive or to amend the listing of a color additive for a new use. The petitioner for a new color additive must provide information on the following:

- Identity of the proposed color additive

- Physical, chemical, and biological properties

- Chemical specifications

- Manufacturing process description

- Stability data

- Intended uses and restrictions

- Labeling

- Tolerances and limitations

- Analytical methods for enforcing chemical specifications

- Analytical methods for determination of the color additive in products

- Identification and determination of any substance formed in or on products because of the use of the color additive

- Safety studies

- Estimate of probable exposure

- Proposed regulation

- Proposed exemption from batch certification

- An environmental assessment or claim for categorical exclusion.

The petitioner must submit data demonstrating the safety and suitability of the new color additive or new use. FDA will then evaluate the data in the petition, public comments to the petition, and other relevant data in FDA's files.

Upon approval of the petition, FDA will issue a new listing regulation or alter an existing regulation for the new color additive or new use. The process for submitting petitions is described in detail in 21 CFR parts 70 and 71, which describe the format, the administrative requirements, and the information and data required. The data that are appropriate for support of a color additive petition will vary depending on whether the petition is for a new color additive or for a new use for a listed color additive, the level and type of use of the proposed color additive, and the amount of color additive and its impurities that may enter body tissues.

Nutraceutical

The term "nutraceutical" is used to describe these medicinally or nutritionally functional foods. Nutraceuticals, which have also been called medical foods, designer foods, phytochemicals, functional foods and nutritional supplements, include such everyday products as "bio" yoghurts and fortified breakfast cereals, as well as vitamins, herbal remedies and even genetically modified foods and supplements. Many different terms and definitions are used in different countries, which can result in confusion.

The term "nutraceutical" was coined in 1989 by Stephen De Felice, founder and chairman of the Foundation for Innovation in Medicine, an American organization which encourages medical health research? He defined a nutraceutical as a "food, or parts of a food, that provide medical or health benefits, including the prevention and treatment of disease".

In Canada, a functional food has been defined as being "similar in appearance to conventional foods consumed as part of a usual diet" whereas a nutraceutical is "a product produced from foods but sold in pills, powders, (potions) and other medicinal forms not generally associated with food".

In Britain, the Ministry of Agriculture, Fisheries and Food has developed a definition of a functional food as "a food that has a component incorporated into it to give it a specific medical or physiological benefit, other than purely nutritional benefit".

Hence, both in Canada and in Britain, a functional food is essentially a food, but a nutraceutical is an isolated or concentrated form. In America, "medical foods" and "dietary supplements" are regulatory terms, however "nutraceuticals", "functional foods", and other such terms are determined by consultants and marketers, based on consumer trends.

Many of these new products that are being promoted to treat various disease states, find their origins in the plant kingdom. This is an obvious choice as many plants produce secondary compounds such as alkaloids to protect themselves from infection and these constituents may be useful in the treatment of human infection. There is also a long history of plant use in many cultures which can be used to identify plants with activity in the treatment of disease.

Dietary Supplements

The term "dietary supplement" describes a broad and diverse category of products that you eat or drink to support good health and supplement the diet. Dietary supplements are not medicines, nor should they be considered a substitute for food.

Dietary ingredients can be one or a combination of any of the following:

- Vitamin

- Mineral

- Herb or other botanical

- Amino acid (the individual building blocks of a protein

- Concentrate, metabolite, constituent, or extract

Although some herbal and mineral compounds have been used for hundreds of years to treat health conditions, today dietary supplement manufacturers are not legally allowed to say their products cure, treat or prevent disease. Supplement makers can say their products support health or contribute to well-being.

That is because Congress does not regulate dietary supplements the same way it regulates medicine. Except for new dietary ingredients, dietary supplement manufacturers do not need to prove to the U.S. Food and Drug Administration (FDA) that a product is safe or effective to be able to sell them. And, unlike medicines, which are required to meet USP standards to help ensure product consistency across multiple manufacturers, USP standards are voluntary for dietary supplements.

Dietary supplements are widely available in health food stores, drug stores, grocery stores, fitness centers and online and they come in many forms including: 2 piece capsules, soft gels, tablets, bottles of liquid, powders and gummies.

Effectiveness

If you don't eat a nutritious variety of foods, some supplements might help you get adequate amounts of essential nutrients. However, supplements can't take the place of the variety of foods that are important to a healthy diet.

Supplements are most likely to cause side effects or harm when people take them instead of prescribed medicines or when people take many supplements in combination. Some supplements can increase the risk of bleeding or, if a person takes them before or after surgery, they can affect the person's response to anesthesia. Dietary supplements can also interact with certain prescription drugs in ways that might cause problems. Here are just a few examples:

- Vitamin K can reduce the ability of the blood thinner Coumadin, to prevent blood from clotting.

- St. John's wort can speed the breakdown of many drugs (including antidepressants and birth control pills) and thereby reduce these drugs' effectiveness.

- Antioxidant supplements, like vitamins C and E, might reduce the effectiveness of some types of cancer chemotherapy.

Keep in mind that some ingredients found in dietary supplements are added to a growing number of foods, including breakfast cereals and beverages. As a result, you may be getting more of these ingredients than you think, and more might not be better. Taking more than you need is always more expensive and can also raise your risk of experiencing side effects. For example, getting too much vitamin A can cause headaches and liver damage, reduce bone strength, and cause birth defects. Excess iron causes nausea and vomiting and may damage the liver and other organs.

Be cautious about taking dietary supplements if you are pregnant or nursing. Also, be careful about giving them (beyond a basic multivitamin/mineral product) to a child. Most dietary supplements have not been well tested for safety in pregnant women, nursing mothers, or children.

Quality

Dietary supplements are complex products. The FDA has established good manufacturing practices (GMPs) for dietary supplements to help ensure their identity, purity, strength, and composition. These GMPs are designed to prevent the inclusion of the wrong ingredient, the addition of too much or too little of an ingredient, the possibility of contamination, and the improper packaging and labeling of a product. The FDA periodically inspects facilities that manufacture dietary supplements.

In addition, several independent organizations offer quality testing and allow products that pass these tests to display their seals of approval. These seals of approval provide assurance that the product was properly manufactured, contains the ingredients listed on the label, and does not contain harmful levels of contaminants. These seals of approval do not guarantee that a product is safe or effective. Organizations that offer this quality testing include:

- U.S. Pharmacopeia

- ConsumerLab.com

- NSF International

Common Misconceptions about Dietary Supplements

Megadosing: The "More is Better" Myth

Many people wonder why dietary supplements like vitamins, herbs, and botanicals are sold without a prescription from a doctor, while medicines (or drugs) are closely regulated and controlled. People often make the mistake of assuming that because supplements are sold over the counter, they are completely safe to take, even in high doses.

In the 1990s there was a trend of "megadosing" antioxidants like vitamin C, beta carotene, and vitamin E. Even though no scientific studies have ever proven that large doses of vitamin C can prevent or cure colds, many people still think this is true. Even now, you may hear claims about other benefits of taking large doses of certain vitamins. But using large doses of vitamins to fight disease in humans is not supported by scientific evidence so far.

In fact, large doses of some vitamins or minerals have been shown to be dangerous and even toxic. For example, too much vitamin C can interfere with the body's ability to absorb copper, a metal that's needed by the body. Too much phosphorous can inhibit the body's absorption of calcium. The body cannot get rid of large doses of vitamins A, D, and K and these can reach toxic levels when too much is taken.

Talk with your doctor before taking large doses of any vitamin, mineral, or other supplement. Your nurse or pharmacist may also be able to give you more information on safe dosages. Even when vitamin doses are not high enough to cause toxic effects, they can have a bad impact on overall health. For instance, several large studies have found that, on average, people taking vitamin E supplements lived no longer than those who didn't. Some even died sooner, particularly of heart failure.

The "Natural is Safe" and "Natural is Better" Myths

In today's world, you won't find much support for the idea that a man-made or refined substance is better or safer than one sold in its unrefined, natural state. But supplements that claim to be "all natural" are not always better or safer than refined or manufactured substances.

Keep in mind that some of the most toxic substances in the world occur naturally. Poison mushrooms, for example, are completely natural but not safe or helpful to humans. Many plants in nature are toxic or deadly if taken internally.

Botanical supplements (such as garlic, ginger, ginkgo biloba, echinacea, and others) are made of plant material, so many of them are sold as "natural" products. But plants are made up of many chemicals. Some of these chemicals can be helpful while others are poisonous or can cause allergies in humans. Botanicals that are marketed as "all natural" are not always the most helpful ones, since they may not be refined to remove potentially harmful chemicals.

Botanicals can contain any or all parts of the plant, including roots, stems, flowers, leaves, pollen, and juices. Different parts of plants can have very different effects on humans. For instance, dandelion root is a laxative (it causes bowel movements), while dandelion leaves contain a diuretic (a chemical that increases urination). If you decide to use a botanical supplement, make sure you know what parts of the plant were put into it. If you're unsure, contact the company and ask them how they make their supplement.

Remember, too, that safety and dose are related. The leaves or roots of many plants can be safely taken in small amounts as an herb. But concentrated extracts sold as liquids or pills may contain the plant's chemicals in far greater amounts and may not be safe.

The "It's been used for Thousands of Years, so it must Work" Myth

Knowing that a botanical has been used in folk or traditional medicine for thousands of years is helpful, but is not convincing proof that it works or that it's safe. If small amounts of a plant caused painful or life-threatening side effects right away, it probably wouldn't have been used in folk medicine or traditional medical systems. But traditional medical systems thousands or even hundreds of years ago did not have the scientific methods to detect long-term side effects. So, if a plant seemed useful over the short term but actually increased the risk of chronic disease (like cancer, heart failure, or kidney failure) after years of use, those side effects would not have been noticed. Also, if a patient's problem got worse after using an herb, the worsening may not have ever been linked to the treatment itself. Deaths weren't unusual; unlike today, people of all ages died of illnesses that can now be prevented, treated, or cured. Finally, in some traditional systems, herbs were given to cause vomiting or diarrhea. These effects may have been considered helpful at the time, even if the final, long-term outcome wasn't good.

It also helps to find out whether a plant used today is being used like it was traditionally. For example, tea prepared from a certain plant might have been safely used in traditional Chinese medicine to treat occasional bouts of asthma when given by an experienced practitioner. On the other hand, daily use of much higher doses taken in a concentrated pill form with no expert supervision might be quite unsafe.

As you consider ancient treatments, remember that most herbs, plants, and other methods were used in traditional medicine systems to reduce symptoms or make the person feel better. This was helpful to people who were likely to recover anyway. Still, it was understood that death was a possible outcome of most serious illnesses. It's safe to say that science and technology have helped us better understand the causes of illness today than anyone did centuries ago. Now, most people whose families once used these traditional healing methods prefer to be treated with modern medicine, if there's a proven treatment available.

The "It can't Hurt to Take Supplements Along with My Regular Medicines" myth

Many people assume that dietary supplements are always safe to take along with prescription drugs. This is not true. For example, certain botanicals can block or speed up the body's absorption of some prescription drugs. This can cause the person to have too much or too little of the prescribed drug in their bloodstream. Most drug companies and producers of herbal supplements do not research possible drug interactions, so the risks of taking supplements with other drugs are largely unknown.

Pink Slime

Pink slime is an epithet for a beef product, where the beef is extracted from cuts of meat by a process that the meat industry calls "lean finely textured beef," abbreviated as LFTB, and "boneless lean beef trimmings," or BLBT. It has been mockingly termed "soylent pink." Pink slime has been claimed by some originally to have been used as pet food and cooking oil and later approved for public consumption, however, both the Food and Drug Administration administrator responsible for approving the product and Beef Products, Inc., the largest U.S. producer, have disputed this. This claim is also one of the subjects of a lawsuit currently before the courts. In 2001, the United States Department of Agriculture approved the product for limited human consumption, and it was used as a food additive to ground beef and beef-based processed meats as a filler, at a ratio of usually no more than 25 percent of any product. The production process uses heat in centrifuges to separate the fat from the meat in beef trimmings. The resulting product is exposed to ammonia gas or citric acid to kill bacteria.

Production and Content

Finely textured meat is produced by heating boneless beef trimmings to 107–109°F (42–43°C), removing the melted fat by centrifugal force using a centrifuge, and flash freezing the remaining product to 15°F (–9°C) in 90 seconds in a roller press freezer. The roller press freezer is a type of freezer that was invented in 1971 by BPI CEO Eldon Roth that can "freeze packages of meat in two minutes" and began to be used at Beef Products Inc. in 1981. The lean finely textured beef is added to ground beef as a filler or to reduce the overall fat content of ground beef. In March 2012 about 70% of ground beef sold in US supermarkets contained the product. Source areas for the product from cattle may include the most contaminated portions, such as near the hide.

The recovered beef material is extruded through long tubes that are thinner than a pencil, during which time at the Beef Products, Inc. (BPI) processing plant, the meat is exposed to gaseous ammonia. At Cargill Meat Solutions, citric acid is used to kill bacteria such as *E. coli* and *Salmonella*.

Gaseous ammonia in contact with the water in the meat produces ammonium hydroxide. The ammonia sharply increases the pH and damages microscopic organisms, the freezing causes ice crystals to form and puncture the organisms' weakened cell walls, and the mechanical stress destroys the organisms altogether. The product is finely ground, compressed into pellets or blocks, flash frozen and then shipped for use as an additive.

Most of the finely textured beef is produced and sold by BPI, Cargill and Tyson Foods.As of March 2012 there was no labeling of the product, and only a USDA Organic label would have indicated that beef contained no "pink slime". Per BPI, the finished product is 94% to 97% lean beef (with a fat content of 3% to 6%) has a nutritional value comparable to 90% lean ground beef, is very high in protein, low in fat, and contains iron, zinc and B vitamins. U.S. beef that contains up to 15% of the product can be labeled as "ground beef". Up to 2005, filler could make up to 25% of ground meat. In an Associated Press review, food editor and cookbook author J.M. Hirsh compared the taste of two burgers: one supposedly containing LFTB and one traditional hamburger. He described the LFTB-containing burgers as smelling the same, but being less juicy and with not as much flavor.

In 2002, a United States Department of Agriculture (USDA) microbiologist stated that the product contained connective tissue and that he did not consider it to be ground beef and that it was "not nutritionally equivalent" to ground beef. Rick Jochum, a spokesperson for BPI, stated in 2012 that BPI's product does not contain cow intestines or connective tissue such as tendons.

Early use

Ground beef that does not contain the LFTB additive

In 1990, the USDA's Food Safety and Inspection Service (FSIS) approved the use of the technology for manufacturing finely textured meat. At the time of its approval, the FSIS called the remaining product "meat", although one FSIS microbiologist dissented, arguing it contained both muscle and connective tissue.

In 1994, in response to public health concerns over pathogenic *E. coli* in beef, the founder of BPI, Eldon Roth, began work on the "pH Enhancement System", which disinfects meat using injected anhydrous ammonia in gaseous form, rapid freezing to 28°F (−2°C), and mechanical stress.

In 2001, the FSIS approved the gaseous disinfection system as an intermediate step before the roller press freezer, and approved the disinfected product for human consumption, as an additive. The FSIS agreed with BPI's suggestion that ammonia was a "processing agent" which did not need to be listed on labels as an ingredient. FSIS microbiologists Carl Custer and Gerald Zirnstein stated that they argued against the product's approval for human consumption, saying that it was not "meat" but actually "salvage", and that the USDA should seek independent verification of its safety, but they were overruled. In 2003, BPI commissioned a study of the effectiveness and safety of the disinfection process; the Iowa State University researchers found no safety concern in the product or in ground beef containing it.

The term "pink slime", a reference to the product's "distinctive look", was coined in 2002 by Zirnstein in an internal FSIS e-mail. Expressing concern that ammonia should be mentioned on the labels of packaged ground beef to which the treated trimmings are added, Zirnstein stated "I do not consider the stuff to be ground beef, and I consider allowing it in ground beef to be a form of fraudulent labeling". He later stated that his main concern was that connective tissue is not "meat", and that ground beef to which the product had been added should not be called ground beef, since it is not nutritionally equivalent to regular ground beef.

In 2007, the USDA determined the disinfection process was so effective that it would be exempt from "routine testing of meat used in hamburger sold to the general public".

A December 2009 investigative piece published by *The New York Times* questioned the safety of the meat treated by this process, pointing to occasions in which process adjustments were not effective. In January 2010, *The New York Times* published an editorial reiterating the concerns posed in the news article while noting that no meat produced by BPI had been linked to any illnesses or outbreaks.

An episode of *Jamie Oliver's Food Revolution* aired on April 12, 2011, depicted Jamie Oliver decrying the use of "pink slime" in the food supply and in school lunches. In the episode, Oliver douses beef trimmings in liquid ammonia while explaining what the product is and why he is disgusted with it. Oliver stated, "Everyone who is told about 'pink slime' doesn't like it in their food—school kids, soldiers, senior citizens all hate it". The introduction of the additive into the nation's meat supply caused concern and was criticized by some scientists. "The scientists said they had used the term 'pink slime' to describe the product, which they said should have been identified as an additive and believed was not actually beef as it is commonly defined." The American Meat Institute and Beef Products Inc. retorted with a YouTube video featuring Dr. Gary Acuff of Texas A&M University questioning some of Oliver's statements and promoting the additive.

ABC News Report

An 11-segment series of reports in March 2012 from ABC News brought widespread public attention to and raised consumer concerns about the product. The product was described as "essentially scrap meat pieces compressed together and treated with an antibacterial agent". Lean finely textured beef (LFTB) was referred to as "an unappetizing example of industrialized food production". The product has been characterized as "unappetizing, but perhaps not more so than other things that are routinely part of hamburger" by Sarah Klein, an attorney for the food safety program at the Center for Science in the Public Interest. Nutritionist Andy Bellatti has referred to the product as "one of many symptoms of a broken food system". Food policy writer Tom Laskawy noted that ammonium hydroxide is only one of several chemicals routinely added to industrially produced meat in the United States.

It was reported at that time that 70% of ground beef sold in US supermarkets contained the additive, and that the USDA considered it as meat. The USDA issued a statement that LFTB was safe and had been included in consumer products for some time, and its Under Secretary of Agriculture for Food Safety Elisabeth A. Hagen stated that "The process used to produce LFTB is safe and has been used for a very long time. And adding LFTB to ground beef does not make that ground beef any less safe to consume".

Industry Response

Manufacturer Beef Products Inc. (BPI) and meat industry organizations addressed public concerns by stating that the additive, though processed, is "lean beef" that simply was not able to be reclaimed through traditional slaughterhouse practices until newer technologies became available approximately 20 years ago. With regard to concerns over the use of ammonium hydroxide, BPI noted that its use as an anti-microbial agent is approved by the Food and Drug Administration. The use of ammonium hydroxide is included on the FDA's list of GRAS (generally recognized as safe) procedures, and is used in similar applications for numerous other food products, including puddings and baked goods.

Market Response

Several U.S. food manufacturers publicly stated that they did not use the product in their wares, including ConAgra Foods Inc., Sara Lee Corporation and Kraft Foods Inc. Many meat retailers stated that they either did not use the product, or would cease using it.

Many fast food chains stopped use of the product after the controversy arose, or stated that they had not used the product before. In April 2012 the *Concord Monitor* reported increased business in some small neighborhood markets where the product's use was less likely, due to consumer concerns about the additive.

On March 25, 2012, BPI announced it would suspend operations at three of its four plants, being in "crisis planning". The three plants produced a total of about 900,000 pounds of the product per day. BPI said it lost contracts with 72 customers, many over the course of one weekend, and production decreased from 5 million pounds of LFTB per week to below one million pounds a week at the nadir (lowest point of production). Effective May 25, 2012 BPI closed three of its four plants, including one in Garden City, Kansas, lost more than $400 million in sales, and laid off 700 workers. Production decreased to less than 2 million pounds in 2013. Cargill also significantly cut production of finely textured beef and in April 2012 "warned that the public's resistance to the filler could lead to higher hamburger prices this barbecue season". About 80% of sales of the product evaporated "overnight" in 2012, per the president of Cargill Beef. Cargill stopped production in Vernon, California, and laid off about 50 workers as well as slowing production at other plants including a beef-processing plant in Plainview, Texas, where about 2,000 people were laid off.

Many grocery stores and supermarkets, including the nation's three largest chains, announced in March 2012 that they would no longer sell products containing the additive. Some grocery companies, restaurants and school districts discontinued the sale and provision of beef containing the additive after the media reports.

In April 2012, the USDA received requests from beef processors to allow voluntary labeling of products with the additive, and stated it planned to approve labeling after checks for label accuracy. Both BPI and Cargill made plans to label products that contain the additive to alleviate these concerns and restore consumer confidence. Following the USDA announcement to allow choices in purchasing decisions for ground beef, many school districts stated that they would opt out of serving ground beef with LFTB. By June 2012, 47 out of 50 U.S. states declined to purchase any of the product for the 2012–2013 school year while South Dakota Department of Education, Nebraska, and Iowa chose to continue buying it.

On April 2, 2012, AFA Foods, a ground-beef processor manufacturer of finely textured beef owned by Yucaipa Companies filed for Chapter 11 bankruptcy citing "ongoing media attention" that has "dramatically reduced the demand for all ground beef products". On April 3, 2012, U.S. cattle futures on the Chicago Mercantile Exchange were at a 3.5-month low, which was partially attributed to the "pink slime" controversy. Livestock traders stated that: "It has put a dent in demand. It is bullish for live cattle over the long-term, but short-term it is certainly negative".

School Lunches

The reaction against the product has also been partially credited to a Change.org petition that has landed over a quarter million signatures to ban it in school lunches. After some parents and consumer advocates insisted the product be removed from public schools, the USDA indicated, beginning in fall 2012, that it would give school districts the choice between ground beef with or without LFTB. CBS News reported that Chicago Public schools may have served "pink slime" in school lunches.

While some school districts have their own suppliers, many school districts purchase beef directly from the USDA and do not know what is in the beef. For the year 2012, the USDA planned on purchasing 7 million pounds of lean beef trimmings for the U.S. national school lunch program. USDA spokesman Mike Jarvis stated that of the 117 million pounds of beef ordered nationally for the school lunch program in the past year, 6% was LFTB. An analysis of California Department of Education data indicated that "anywhere from none to nearly 3 million pounds of beef from the USDA that was served in California schools last year could have contained lean finely textured beef". According to the USDA, the cost differential between ground beef with and without the additive has been estimated at approximately 3%.

Current use

In March 2012, 70% of ground beef in the U.S. contained lean finely textured beef, and a year later in March 2013 the amount was estimated by meat industry officials to be at approximately 5%. This significant reduction is due in part to the extensive media coverage that began in March 2012 about the additive. Kroger Co. and Supervalu Inc. have stopped using the additive.

Cargill started using a label stating "Contains Finely Textured Beef" from 2014. Production of finely textured beef increased modestly, as beef prices rose by 27% over two years in 2014 and "retailers sought cheaper trimmings to include in hamburger meat and processors find new products to put it in". Senior management of Cargill claimed almost full recovery as sales tripled. BPI regained 40 customers that are mostly processors and patty-makers who distribute to retailers and the USDA since March 2012. It does not label its product.

Regulation

Iowa governor Terry Branstad, a supporter of the product's use in beef products

In the US, the additive is not for direct consumer sale. Lean finely textured beef can constitute up to 15% of ground beef without additional labeling, and it can be added to other meat products such as beef-based processed meats.

Because of ammonium hydroxide use in its processing, the lean finely textured beef by BPI is not permitted in Canada. Health Canada stated that: "Ammonia is not permitted in Canada to be used in ground beef or meats during their production" and may not be imported, as the Canadian Food and Drugs Act requires that imported meat products meet the same standards and requirements as domestic meat. Canada does allow Cargill's citric acid-produced Finely Textured Meat (FTM) to be "used in the preparation of ground meat" and "identified as ground meat" under certain conditions.

Lean finely textured beef and Finely Textured Meat is banned for human consumption in the European Union (EU).

Public Perception

The nature of the product and the manner in which it is processed led to concerns that it might be a risk to human health, despite the fact that there have been no reported cases of foodborne illnesses due to consumption of the product. Among consumers, media reporting significantly reduced its acceptance as an additive to ground beef.

A Harris Interactive survey commissioned by Red Robin and released on April 4, 2012, found that 88% of US adults were aware of the "pink slime" issue, and that of those who were aware, 76% indicated that they were "at least somewhat concerned", with 30% "extremely concerned". 53% of respondents who stated that they were aware of pink slime took some action, such as researching ground beef they purchase or consume, or decreasing or eliminating ground beef consumption.

Food Coating

Coating is a major stage in food processing controlling a product's development, arrangement, texture and taste structures. The procedure includes placing the product particles in motion and instantaneously applying the coating ingredient in a particular pattern to expose one to another. Thus, coated food ingredients are delivering unique possibilities for development of product, quality and processing enhancement. Escalating food processing industry will certainly drive the Global Food Coating Ingredients market in the coming few years.

A coating influences your yield and will largely determine the bite, taste and presentation of your end products. Presentation plays a key factor in the consumer's choice for a product. It is of vital importance to get your coating right.

Coating enhances taste and texture and increases yield. It also protects products from damage during freezing. With a coating, your products look good. The number of possible coatings is abundant. Will your choice be for wet or dry, or rather a combination of both? The process sequence will depend on the end product you choose. Regardless of which coating you choose, Marel is able to supply the appropriate system for virtually all types of coating.

- Coatings for food products

    Understanding the intent of a coating is important; agreeing on the scope helps in opening the research field, selecting options, and making decisions at key milestones of the project.

    Food products come in an infinite variety. Many of them are coated to convey an enticing aspect, palatability, enjoyment or protection. Knowing the product's characteristics is essential to the fulfilment of the purpose.

- Coating ingredients

Traditionally, coating ingredients were mostly derived from nature and minimally processed: honey, sugar, chocolate, sauce, etc. There were few choices but significant quality variations.

Nowadays, the range of coating ingredients has grown tremendously and they are created to serve specific purposes. A slight change in the choice of the ingredient may greatly help the whole process in terms of feasibility and simplification. At the same time, the tolerance in their characteristics has narrowed in order to offer reliable and repeatable behaviour during the production process.

- The food coating process

A coating process fulfils two basic functions – application and motion – which can be further split into multiple sub-operations. Application implies storing, dosing, heating or cooling, and dispersing the ingredient. Moving the product means creating the right motion to expose it to the ingredient. In addition, post treatment (drying, cooling, freezing and polymerizing) also has a significant influence on the end result.

- Coating machines

The coating process is implemented by a complex assembly of dedicated equipment: conveyor, tumbler, feeder, pump, etc. Although base designs readily exist on the market, they often need to be adapted to cope with the required specifications.

The installed process is expected to meet the required quality (function, dosage, homogeneity) and quantity (capacity, saving). Because of the nature of the process, it faces specific challenges in minimising the causes of downtime; e.g. clogging, pollution, cleaning and recipe changeover time.

References

- Zhou, Shi-Sheng (2014). "Excess vitamin intake: An unrecognized risk factor for obesity". World J Diabetes. 5 (1): 1–13. doi:10.4239/wjd.v5.i1.1. ISSN 1948-9358. PMC 3932423. PMID 24567797

- Food-additive: britannica.com, Retrieved 18 June 2018

- Sangani, Rahul; Ghio, Andrew (2013). "Iron, Human Growth, and the Global Epidemic of Obesity". Nutrients. 5 (10): 4231–4249. doi:10.3390/nu5104231. ISSN 2072-6643. PMC 3820071. PMID 24152754

- What-is-food-coloring: naturallysavvy.com, Retrieved 12 July 2018

- Richardson, D. P. (28 February 2007). "Food Fortification". Proceedings of the Nutrition Society. 49 (1): 39–50. doi:10.1079/PNS19900007

- Regulatory process historical perspectives, coloradditives, forindustry: fda.gov, Retrieved 28 May 2018

- Park YK, Sempos CT, Barton CN, Vanderveen JE, Yetley EA (2000). "Effectiveness of food fortification in the United States: the case of pellagra". American Journal of Public Health. 90 (5): 727–38. doi:10.2105/AJPH.90.5.727. PMC 1446222. PMID 10800421

- What-dietary-supplement: quality-supplements.org, Retrieved 18 March 2018

- Özer, Cem Okan; Kılıç, Birol. "New Discussion Subject of Meat Industry: "Pink Slime"". Turkish Journal of Agriculture. 2 (6 (2014)). ISSN 2148-127X. Retrieved December 9, 2014

- Food-coating-ingredients-market: futuremarketinsights.com, Retrieved 30 March 2018

- Irwin H. Rosenberg (August 2005). "Science-based micronutrient fortification: which nutrients, how much, and how to know?". The American Journal of Clinical Nutrition. 82(2): 279–280. PMID 16087969

Industrial Methods of Food Production

Science and technology have undergone rapid developments in the past decade, which has resulted in innovation in industrial methods of food production such as snap freezing, ultra-high temperature processing, dry milling and fractionation of grain, Ohmic heating, etc. These have been extensively discussed in this chapter. It also explores the principles of basic industrial processes like food drying, brining and extrusion for a comprehensive understanding of the field.

Brining

Brining foods in a saltwater mixture before you cook them adds flavor, tenderness, and reduces cooking times. If this sounds like a good thing, then it's time to learn the basics about brining.

Brining meat is an age-old process of food preservation. Heavy concentrations of salt-preserved meats were taken on long ocean voyages and military campaigns before the advent of refrigeration. Today, brining has a new purpose. By using smaller quantities of salt mixed with other spices and herbs, brining can permeate meat with flavor.

Working of Brining

The chemistry behind brining is pretty simple. Meat already contains salt water. By immersing meats in a liquid with a higher concentration of salt, the brine is absorbed into the meat. Any flavoring added to the brine will be carried into the meat with the saltwater mixture. Because the meat is now loaded with extra moisture, it will stay that way as it cooks.

The process of brining is easy but takes some planning. Depending on the size of what you want to brine it can take up to 24 hours or more. If you are going to be brining a whole bird, you will also want an additional 6 to 12 hours between the brining and the cooking. If you want your poultry to have a golden, crispy skin, it needs to sit in the refrigerator for several hours after you remove it from the brine so that the meat can absorb the moisture from the skin.

The most basic process of brining is to take approximately 1 cup of table salt (no iodine or other additives) to 1 gallon of water. Another way to measure this concentration is with a raw egg. The ideal brine has enough salt to float a raw egg. You will need enough brine to completely submerge the meat without any part being out of the liquid.

Some items might need to be weighed down to stay under. Brine meat for about an hour per pound. Remove from brine (don't reuse the brine) and rinse to remove any excess salt before cooking.

Process of Making a Basic Brine

The typical brine consists of 1 cup of salt for each gallon of water (or other liquids). Start by determining the amount of liquid you are going to need. To do this take the meat you plan to brine and place it in the container you are going to use. The container can be most anything that will easily fit the meat but isn't so big that you have to prepare far more brine that you need. Plastic containers, crocks, stainless steel bowls, resealable bags or any non-corrosive material will work.

Once you know how much liquid is needed start by boiling 2 cups of water for each cup of salt, you will need. Once it boils, add the salt (and sugar if you are going to be using sugar) and stir until dissolved.

Add other spices and herbs. Combine with the remaining liquid (should be cold). The brine should always be cold before you add the meat so you should refrigerate it before you add the meat. You don't want the brine cooking the meat.

At this point, you can add other brine ingredients like juices or cut up fruit. Submerge the meat into the brine. You can use a plate or other heavy object to keep it down. It is important that no part of the meat be exposed to the air. Saltwater brines will kill bacteria and keep the meat from spoiling, but it doesn't work if part of the meat is sticking out.

Brine meats for about 1 hour per pound in the refrigerator. It is important that everything is kept cold. The specific amount of time will vary. Lighter meats like poultry or seafood do not need to be brined as long as denser meats like pork tenderloins.

Use the following chart to give you an idea of how long to brine. Remember that the longer you brine, the stronger the flavor will be. If you over brine you could end up with some very salty meat.

Once the meat is properly brined, remove it. You do not need to rinse unless you were using a high salt concentration in the brine or if there is a layer of visible salt on the surface. Otherwise, you can take cuts of meat straight to the grill, smoker, or oven. Whole poultry is the exception, however. To get a crispy, brown skin, whole birds should be removed from the brine, wrapped in foil or plastic and put in the refrigerator overnight or for at least 12 hours.

Basic Brining Times

Meat	Brine Time
Shrimp	30 Minutes
Whole Chicken (4 to 5 pounds)	4 to 5 hours
Turkey (12 to 14 pounds)	12 hours
Pork Tenderloin (whole)	12 hours
Cornish Hens	1 to 2 hours

Food Drying

Drying is the simple process of dehydrating foods until there is not enough moisture to support microbial activity. Drying removes the water needed by bacteria, yeasts, and molds need to grow. If adequately dried and properly stored, dehydrated foods are shelf stable (safe for storage at room temperature). The drying food preservation method is easy to do, very safe, and can be used for most types of foods (meats, fruits, and vegetables).

There are several methods for drying foods. Two of the easiest and most common that can be used in any climate are oven drying and drying with an electric dehydrator appliance; these methods are described below. The other methods are air drying (in the shade during warm weather), sun drying (limited to desert climates), solar drying (requires specially built dryer), and pit oven drying (useful when other methods are impractical). Find these other food drying methods describe in The Home Preserving Bible by Carole Cancler.

Process to Dry Food in a Conventional Oven

Oven drying is a good choice if you have never dried foods before, or plan to do only occasional drying. It tends to be slower than an electric dehydrator, but there is little or no investment in equipment and you don't have to depend on the weather as with other methods.

Foods that are well-suited to oven drying are meats; seafood; fruit leather; low-moisture foods such as herbs, potatoes, bread cubes, berries, and meaty tomatoes (roma or paste-type); and excess produce you might otherwise throw out, such as onions, celery, and bananas. If you are new to drying, start with a few of the easiest foods to dry: berries, banana slices, tomato slices, chopped onions, oven jerky, and smoked salmon.

Here are the basic steps for oven drying foods:

- Prepare suitable trays for drying foods.

- Prepare food for drying: Preparation methods vary depending on the food you want to dry. For fruits and vegetables, you wash and then usually halve, quarter, or slice the produce. For light colored fruits and all vegetables, you also steam-blanch to deactivate enzymes or prevent browning in light colored foods, and then pat dry.

- Preheat (a gas or electric) oven: to the lowest temperature setting. Maintain an oven temperature between 125°F and 145°F. Check the oven temperature with an accurate thermometer.

- Decrease the temperature: by propping open the oven door with a wooden spoon or folded towel.

- Caution: the oven-drying method is not safe in a home with small children.

- Maximize air circulation: to speed drying. Place a fan on a chair near the propped-open oven door so that it blows away the hot, escaping air. Open nearby doors and windows to promote more airflow.

- Dry until pliable or crisp: The extent of dryness is somewhat a matter of preference Therefore, the length of drying time can fluctuate widely (from a few hours to more than 24). Drying time also depends on several factors: the type of food (meat, fruit, vegetables, etc.), the size of the portions to be dried (thick or thin), the drying method used (sun, air, oven), and the weather (especially humidity, which greatly increases drying time).

- Tips for successful drying: include drying foods only on days when the humidity is not high, space the food about an inch apart, and fill only half of the oven racks with food.

Suitable Trays for Oven-drying Foods

Trays used for drying foods in an oven (or other methods than a food dehydrator) need to be of a food-safe screen material such as plastic (preferably polypropylene), stainless steel, Teflon or Teflon coated fiberglass, or wood. An economical solution is to stretch cheesecloth or natural muslin over an oven or cake rack or a wood frame, and attach it with masking tape, paper clips, or clothespins. For a more permanent, but more costly option, have (window) screens made at a hardware store and use them for drying.

Avoid materials which can leach harmful chemicals, darken the food, or melt at drying temperatures. These materials include:

- Do not use uncoated fiberglass and vinyl.

- Do not use metals other than stainless steel (such as aluminum, galvanized steel, and copper); they can transfer a metallic flavor to food, rendering it inedible. Covering metal with cheesecloth or muslin is another option, especially if you are re-purposing material and are unsure of the type of metal.

- Do not use green wood, pine, cedar, oak, and redwood.

After oven drying a few foods, if you want to continue to use the drying method, consider investing in an electric food dehydrator.

Using of Dried Foods

You can used dried foods in a variety of ways:

- Eat dried foods as is (such as snacking on dried beef jerky and dried fruits);

- Rehydrate dried foods water (such as adding vegetables to a meat stew);

- Grind dried foods into a powder (for example, grind tomatoes to a powder that you can reconstitute with water to make tomato sauce).

Therefore, you may dry foods until pliable, especially if you want to use them as a snack food. If you want to store dried food longer or use it to grind to a powder (such as tomatoes to make sauce), then you want them to be crisp and brittle. Less-dry products have considerably shorter shelf life—from 2 weeks to 2 months. Very dry foods, if properly stored, may last several months.

Whether pliable or crisp, condition all foods at the end of the drying process. Alternatively, you may store partially dried or unconditioned foods in the freezer.

How to Condition and Store Foods after Drying

Individual pieces of food dry at different rates; some pieces will have more moisture than others. If there is too much moisture left in a few pieces, they can grow mold and contaminate the entire batch. To guard against mold growth, you need to condition dried foods before you store them. During conditioning, the moisture will equalize—that is, excess moisture will transfer to drier pieces, until it is evenly distributed throughout the batch.

To condition dried foods, place them in a tightly closed container at room temperature. Stir or shake the contents every day for a week. If you open the container to stir the contents, be sure to close it again tightly. During conditioning, if moisture forms on the inside of the container, the food is not sufficiently dry and you need to return it to the dryer.

To store dried foods after conditioning, seal dried food in airtight containers that hold only enough food to be used at one time. This reduces the number of times a package is reopened. You can also limit air by taping over jar enclosures or using a desiccant to absorb oxygen. Ideally, you want to store dried foods at a constant temperature between 40°F and 70°F. Be sure to store foods in a closed cupboard or dark room, away from light. If you live in a dry climate, your dehydrated foods will tend to stay fresh longer. However, if you live in a humid area, moisture can get in and shorten storage life considerably. In high-humidity locations, put dried food in zipper-lock plastic bags that allow you to push out excess air.

You can store properly packaged, well-dried foods at room temperature for up to 1 year. Less dry, pliable products have a shelf life of a few weeks to several months. Storage life decreases with packaging that is not airtight, reopening packages, and fluctuating temperatures. You can vacuum-seal, refrigerate, or freeze any dried food for longer storage.

Check dried foods monthly for spoilage—usually mold. Use dried foods before other types of preserved foods, such as frozen or canned. Most importantly, enjoy eating your dried foods and be sure to experiment with different ways of using your stored treasures.

Tips for Successfully Storing Dried Foods

- Always store dried foods in airtight containers in a cool, dry place.

- Reduce the number of times a package is reopened by using containers that hold only enough food to be used at one time.

- Limit air, light, and heat. Put masking tape over jar enclosures or use a food-safe desiccant in the jar to absorb excess oxygen. Be sure to store foods in a closed cupboard or dark room, away from light. Ideally, you want to store dried foods at a constant temperature between 40°F and 70°F.

- In humid locations, put dried food in zipper-lock plastic bags that allow you to push out excess air. This helps to prevent moisture from re-entering the food, shortening the storage life considerably.

- To increase storage life, vacuum-seal, refrigerate, or freeze dried foods.

Food Extrusion

In the modern world of globalization, consumers are looking forward to a healthy nutrition at an affordable price. Extrusion systems meet the current, constantly changing requirements in a very flexible and economical manner. Food Extrusion plays an important role in the manufacture of pasta, ready-to-eat cereals, snacks and pet foods. The process involves a combination of several unit operations including mixing, cooking, kneading, shearing, shaping and forming. Food Extrusion cooking has gained popularity over the last two decades for a number of reasons:

- Versatility

- Cost

- Productivity

- Product Quality

- Environment-friendly

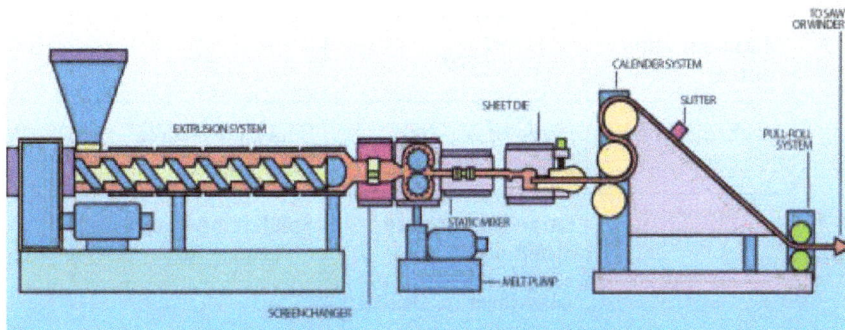

Processes involved in Food Extrusion

Principle

The principles of operation are similar in all types: raw materials are fed into the extruder barrel and the screw(s) then convey the food to it. Further down the barrel, smaller flights restrict the volume and increase the resistance to movement of the food. As a result, it fills the barrel and the spaces between the screw flights and becomes compressed.

As it moves further along the barrel, the screw kneads the material into a semi-solid, plasticized mass. Hot extrusion involves heating above 100°C. Here, frictional heat and any additional heating cause temperature to rise rapidly. Further, the smaller flight section of the barrel increases pressure and shearing.

Finally, it is forced through one or more restricted openings (dies) at the discharge end of the barrel. As the food emerges under pressure from the die, it expands to the final shape and cools rapidly flashing off moisture as steam . A variety of shapes, rods, spheres, doughnuts, tubes, strips or shells can be formed. Typical products include a wide variety of low density, expanded snack foods and ready-to-eat (RTE) puffed cereals.

For pasta and meat products, an implication of cold extrusion technique which involves ambient temperature to mix and shape food. Low-pressure extrusion, at temperatures below 100°C, is used to produce, for example, liquorice, fish pastes, surimi and pet foods.

Extrusion cooking involves high-temperature-short-time (HTST) process which reduces microbial contamination and inactivates enzymes. The main method of preservation of both hot- and cold-extruded foods is by the low water activity of the product (0.1–0.4), and for semi-moist products in particular, by the packaging materials that are used.

Classification of Extruders

- Extruders are classified into two types according to operation: Hot and cold extruders

- Based on type of construction: Single screw and twin screw extruder

Parameters	Single Screw	Twin Screw
Transport Mechanism	Friction between metal and food material	Positive displacement
Capital cost	low	High(twice)
Minimum water content	10%	8%
Mechanical power dissipation	Large shear forces (550-6000kpa)	Small shear force (2000-4000)
Heat distribution	Large temperature difference	Small temperature difference

Classification Of Extruders

Physiochemical Changes during Food Extrusion

Major changes occurs during extrusion process are:

Changes in Starches

The major difference between extrusion processing and conventional food processing is that in the former starch gelatinization occurs at much lower moisture content(12-22%). Once inside the extruder, and at relatively high temperatures, the starch granules melt and become soft, besides changing their structure compressing to a flattened form. The application of heat, the action of shear on the starch granule and water content destroy the organized molecular structure, also resulting in molecular hydrolysis of the material. The starch polymers are then dispersed and degraded to form a continuous fluid melt. The fluid polymer continuum retains water vapour bubbles and stretches during extrudate expansion until the rupture of cell structure. The starch polymer cell walls recoil and stiffen as they cool to stabilize the extrudate structure. Finally, the starch polymer becomes glassy as moisture is removed, forming a hard brittle texture.

Changes in Protein

Proteins are biopolymers with a great number of chemical groups when compared to polysaccharides and are therefore more reactive and undergo many changes during the extrusion process, with the most important being denaturation. Electrostatic and hydrophobic interactions favour the formation of insoluble aggregates. The creation of new peptide bonds during extrusion is controversial. High molecular weight proteins can dissociate into smaller subunits. Enzymes, also proteins, lose their activity after being submitted to the extrusion process due to high temperatures and shear.

Changes in Lipids

Fats and oils can be described as lipids. Lipids have a powerful influence in extrusion cooking processes by acting as lubricants because they reduce the friction between particles in the mix and between the screw and barrel surfaces and the fluid melt.

Changes in Fibres

Research has shown that cooking fibres by extrusion can produce changes in their structural characteristics and physicochemical properties, with the main effect being a redistribution of insoluble fibre to soluble fibre. This effect would be the result of the rupture of covalent and noncovalent bonds between carbohydrates and proteins associated to the fibre, resulting in smaller molecular fragments, that would be more soluble.

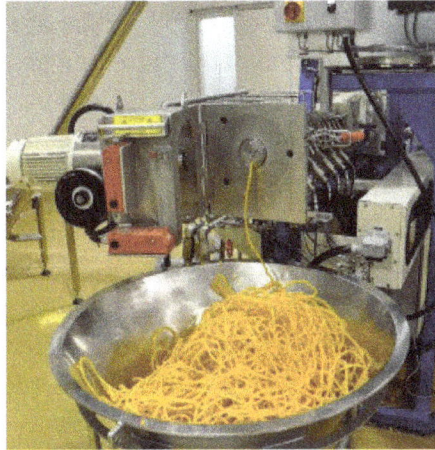

Effects

Extrusion enables mass production of food via a continuous, efficient system that ensures uniformity of the final product. This is achieved by controlling various aspects of the extrusion process. It has also enabled the production of new processed food products and "revolutionized many conventional snack manufacturing processes". The extrusion process results in "chemical reactions that occur within the extruder barrel and at the die". Extrusion has the following effects:

- Destruction of certain naturally occurring toxins.

- Reduction of microorganisms in the final product.

- Slight increase of iron-bioavailability.

- Creation of insulin-desensitizing starches (a potential risk-factor for developing diabetes).

- Loss of lysine, an essential amino acid necessary for developmental growth and nitrogen management.

- Simplification of complex starches, increasing rates of tooth decay.

- Increase of glycemic index of the processed food, as the "extrusion process significantly increased the availability of carbohydrates for digestion".

- Destruction of Vitamin A (beta-carotene).

- Denaturation of proteins.

The material of which an extrusion die is made can affect the final product. Compared to stainless steel dies, a pasta machine with bronze dies produces a rougher surface. This is considered to give an improved taste, as it better retains pasta sauces. "Bronze die" pasta is labelled as such on retail packages, to indicate a premium product.

The effects of "extrusion cooking on nutritional quality are ambiguous", as extrusion may change carbohydrates, dietary fibre, the protein and amino acid profile, vitamins, and mineral content of the extrudate in a manner that is beneficial or harmful.

High-temperature extrusion for a short duration "minimizes losses in vitamins and amino acids". Extrusion enables mass production of some food, and will "denature anti nutritional factors", such as destroying toxins or killing microorganisms. It may also improve "protein quality and digestibility", and affects the product's shape, texture, colour, and flavour.

It may also cause the fragmentation of proteins, starches, and non-starch polysaccharides to create "reactive molecules that may form new linkages not found in nature". This includes Maillard reactions which reduce the nutritional value of the proteins. Vitamins with heat lability may be destroyed. As of 1998, little is known about the stability or bioavailability of phytochemicals involved in extrusion. Nutritional quality has been found to improve with moderate conditions (short duration, high moisture, low temperature), whereas a negative effect on nutritional quality of the extrudate occurs with a high temperature (at least 200°C), low moisture (less than 15%), or improper components in the mix.

A 2012 research paper indicates that use of non-traditional cereal flours, such as amaranth, buckwheat or millet, may be used to reduce the glycemic index of breakfast cereals produced by extrusion. The extrudate using these cereal flours exhibits a higher bulk and product density, had a similar expansion ratio, and had "a significant reduction in readily digestible carbohydrates and slowly digestible carbohydrates". A 2008 paper states that replacing 5% to 15% of the wheat flour and white flour with dietary fibre in the extrudate breakfast cereal mix significantly reduces "the rate and extent of carbohydrate hydrolysis of the extruded products", which increased the level of slowly digested carbohydrates and reduced the level of quickly digested carbohydrates.

Products

Extrusion has enabled the production of new processed food products and "revolutionized many conventional snack manufacturing processes".

Cheese curls made using an extruder

The various types of food products manufactured by extrusion typically have a high starch content. *Directly expanded* types include breakfast cereals and corn curls, and are made in high temperature, low moisture conditions under high shear. *Unexpanded* products include pasta, which is produced at intermediate moisture (about 40%) and low temperature. *Texturized* products include meat analogues, which are made using plant proteins ("textured vegetable protein") and a long die to "impart a fibrous, meat-like structure to the extrudate", and fish paste. Confectionery made via extrusion includes chewing gum, liquorice, and toffee.

Some processed cheeses and cheese analogues are also made by extrusion. Processed cheeses extruded with low moisture and temperature "might be better suited for manufacturing using extrusion technology" than those at high moisture or temperature. Lower moisture cheeses are firmer and chewier, and cheddar cheese with low moisture and an extrusion temperature of 80 °C was preferred by subjects in a study to other extruded cheddar cheese produced under different conditions. An extrudate mean residence time of about 100 seconds can produce "processed cheeses or cheese analogues of varying texture (spreadable to sliceable)".

Other food products often produced by extrusion include some breads (croutons, bread sticks, and flat breads), various ready-to-eat snacks, pre-made cookie dough, some baby foods, some beverages, and dry and semi-moist pet foods. Specific examples include cheese curls, macaroni, Fig Newtons, jelly beans, sevai, and some french fries. Extrusion is also used to modify starch and to pellet animal feed.

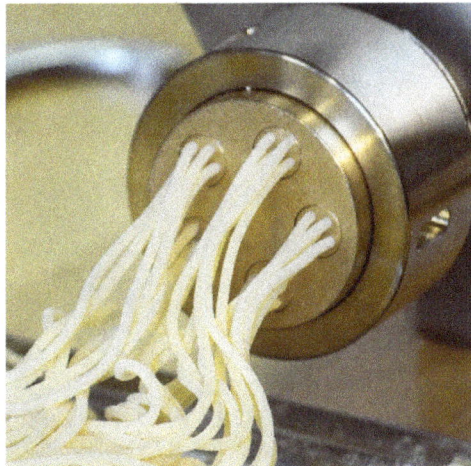

Mass Production

Thousands of companies use mass production every day to get their products onto store shelves. Learn the definition of mass production and go over some techniques and examples.

Take a second to look around your home. What do you see? Furniture? Books? Boxes of food? No matter what your eyes fall on, it's more than likely that several things in your home have been mass produced.

Mass production is a method of production that uses a standardized process of creating interchangeable parts in large quantities for a low price. In other words, a standard process for making products is repeated so each time the product is finished it is exactly the same as all the other parts. The parts are then said to be interchangeable. Whether you use the first part created or the millionth part, they should be exactly the same, with no variation in the outcome.

Mass production decreases the amount of time that workers spend on each individual product. This can allow manufacturers to lower the cost per unit--if the production of each unit is cheaper, its price can be as well. With mass production, manufacturers are also able to increase the amount of units produced and are able to control product quality because the process of producing the good does not vary.

The benefits of mass production include:

- Lower cost per unit

- Decreased time producing products

- Increased output

- Quality control

In this age of mass consumption, many products that we use on a daily basis are mass produced. It could be anything from the furniture you are sitting on to the toys you played with as a kid. The mass production of goods depends on the mass consumption of goods. As the number of consumers and their need for goods increases, it becomes necessary to adopt practices that increase production.

Mass-production is central to fast food. It allows restaurants to receive and store a large amount of food, then cook and serve it as needed. It also allows the companies that own the restaurants' name and trademarks to control what the food looks, smells and tastes like. But, in general, mass-produced fast food is a little different from similar dishes prepared at home:

- It's higher in fat and lower in fiber

- It's higher in calories and sugar

- It's higher in salt

A Big Mac, large fries, baked apple pie and large Coke

In fact, according to McDonald's nutrition charts, if you eat a Big Mac, a large order of fries and a baked apple pie and drink a large Coke, you'll be consuming:

- 1701 calories
- 72 grams of fat
- 1630 milligrams of sodium

The United States Department of Agriculture (USDA) Food Guide recommends that most adults limit their daily consumption to about:

- 2000 calories
- 65 grams of fat
- 1779 milligrams of sodium

A regular burger, small fries and small Coke

In other words, in one fast-food meal, you can consume almost as many calories and sodium and more fat than you should consume in a day. Even if you have a relatively small meal, like a burger, small fries and a small Coke (no dessert), you'll still consume 640 calories, 20 grams of fat and 700 milligrams of sodium. But most fast-food menus are built around large portions. Studies suggest that the size of the portions and the fact that the next larger size only costs a little more encourage people to eat a lot more than they normally would.

In addition, many of the fats in most American fast foods are trans fats, or partially hydrogenated oils. These oils taste good but increase people's risk of heart disease more than other fats do. Researchers believe that consuming more than 5 grams of trans fats per day increases a person's risk of heart attack by 25 percent. More than half of the large-sized fast-food meals analyzed in one study surpassed the 5-gram limit.

Studies on monkeys also suggest that trans fats cause people to gain weight faster than other fats. In one 6-year study, monkeys whose diet included trans fats gained 7.2 percent of their body weight. Other monkeys who consumed the same amount of total fat - but no trans fat - gained only 1.8 percent of their body weight. Some chains have even been sued for the trans fat content of their food. However, people like the taste of trans fats, and they're cheaper than other oils.

The fat, salt and trans fat are part of the allure of fast food - they taste good. Some researchers also theorize that fast food's fat content and its ability to deliver lots of calories in one sitting give it an addictive quality. In addition, the high concentrations of salt, fat and sugar are necessary to make mass-produced food flavorful without making it too expensive.

The mass-production process involves sending lots of food through a factory in a short amount of time. For this reason, one sick animal or piece of waste can contaminate a large amount of food very quickly. Bacteria, viruses and parasites can spread to all of the food in a large mixing vat, or it can contaminate machinery used to process the food.

Snap Freezing

Snap freezing is a terrific way to preserve vegetables, as it keeps for up to 12 months in a correctly maintained freezer, you can cook the vegetables from frozen, it allows purchasing vegetables in bulk to save money, always have vegetables on hand for meals, high level of nutrients is maintained, you can fully understand where it comes from and there is an easy method to follow that uses minimal equipment.

This method applies to vegetables such as carrots, peas, corn, broccoli, cauliflower, turnips, parsnips, onion, snow peas, potatoes, pumpkin, capsicum beans and celery. It is not recommended for lettuce or Asian green leaves. Spinach can use this method, but drain really well and freeze into a block.

This is applicable for vegetables that will be cooked prior to serving, it's not recommended for salads or raw, as they have been blanched in the process of preserving.

Steps of Snap Freeze

Step one: Wash and peel then cut the vegetables into inch cubes.

Step two: Submerge the vegetables in unsalted boiling water. On a high heat, bring the water back to the boil and allow to 'cook' for 1 minute. Then drain really well.

Step three: Separate the pieces on a tray lined with greaseproof paper. Then leave on the bench until it has completely stopped steaming.

Step four: Place the tray in the freezer.

Step five: Once frozen solid, place them in zip-lock bags, removing as much air as possible before sealing. Label and place back into the freezer.

Further Explanation

Cutting into small pieces is easier to freeze. The pieces are already to portion size so it's easier to serve.

Blanching is the process of submerging a food item into boiling water. The purpose is to remove impurities. The water should not be salted for this method.

Allowing time to stop steaming before freezing, makes sure that is doesn't form freezer / ice burn. This will damage the food product and reduce it's quality.

Freezing the items separately means they stay separate until defrosted. It's exactly how the larger companies 'snap freeze' peas and other vegetables. So you can take out the amount you require without defrosting the whole package.

Zip-lock bags ideal for the following reasons:

1. Can make the package almost airtight when sealed.

2. It is resealable.

3. It can label them with a permanent marker.

4. The bags can stack quite easily in the freezer.

5. The bags come in different sizes and are very affordable.

Ultra-high Temperature Processing

The UHT treatment was developed to minimize damage to milk components caused by in-container sterilization while killing or inactivating all microorganisms, so there is little likelihood of microorganisms spoiling the product during storage and transport. The product is said to be 'commercially sterile.' The term 'UHT' may stand for ultra heat treatment or ultrahigh temperature.

UHT processing is being used increasingly for heat-treating milk, as the shelf-life is extended from days to months, and UHT milk can be stored and transported without refrigeration. The use of UHT treatment varies around the world. In Germany, France, and Spain, more than 50% of liquid milk is consumed in UHT form, whereas the process is rarely used in the USA.

The taste and appearance of UHT-treated milk are different from those of pasteurized milk. The milk may taste cooked or burnt to some consumers, because of the development of a sulfurous flavor during the heat treatment. The form of heat treatment used influences the intensity of the sulfurous flavor. However, the flavor has generally dissipated by the time of consumption, and the milk has a slightly brown color.

In some countries, the temperature and time conditions for UHT treatment are specified by regulation. Typical temperature–time conditions for UHT treatment of milk are 130–150°C for 1–3 s. Products with increased solids will require a higher temperature or longer time to ensure effective heat treatment.

UHT treatment can be carried out using a number of different systems. These include direct means of heat treatment such as injection of steam into milk and infusion of milk into a steam chamber or indirect means using heat exchangers where the milk is separated from the heating medium. Scraped-surface heat exchangers may be necessary for UHT treatment of viscous products or products containing particulate matter. When steam is used for direct heat treatment, care must be taken to ensure that there is no dilution of the milk. The steam must also be suitable for contact with food products. All UHT equipment must be sterile to prevent recontamination of the milk with microorganisms. Products that have been UHT-treated must be packed aseptically to ensure optimum shelf-life.

The raw milk to be used for UHT treatment must be selected carefully to ensure that there is minimal chance of heat-stable indigenous or microbial enzymes present causing gelation in the stored products.

Some dairies use equipment designed for UHT treatment to heat the product to 120–130°C without a holding time. This is claimed to double the shelf-life of product packed under non aseptic conditions compared with normal pasteurized milk.

Microbiological Aspects

The main microbiological aim of UHT processing is to inactivate spore-forming bacteria which could grow during storage and cause spoilage. The main targets are Bacillus species, particularly heat-resistant ones such as B. licheniformis and B. subtilus. Geobacillus stearothermophilus is an extremely heat-resistant sporeformer found in milk but because it only grows at temperatures above about 50°C, it does not cause problems in UHT milk unless the milk is severely temperature abused during storage. In comparatively recent times, another extremely heat-resistant sporeformer, B. sporothermodurans, has caused problems in UHT milk; unfortunately, unlike G. stearothermophilus, this organism is mesophilic, that is, it can grow at room temperature.

The thermal conditions used for UHT processing are designed to give a 9-log reduction in heat-resistant sporeformers. This is equivalent to a bacteriological index (B*) of 1. A UHT process should have a B* of at least 1. In practice, most UHT plants exceed this requirement by a reasonable margin. Many different temperature-time combinations ranging from 130°C for ~30 seconds to 160°C for less than 0.05 seconds could achieve this goal but in reality very long and very short holding times are not commercially practical. The most common UHT temperatures used commercially range from 137 to 145°C.

The UHT Process

The major steps in a UHT process are as follows:

- Preheating, with or without a holding time
- Homogenisation (for indirect systems)
- Heating to sterilisation temperature
- Holding at sterilisation temperature
- Initial cooling
- Homogenisation (alternative position for direct or indirect systems)
- Final cooling
- Aseptic packaging

The preheating stage takes the temperature from ~ 5°C to ~90°C, using the hot milk post-sterilisation as the heating source in tubular or plate heat exchangers. This heat regeneration step is very important for the energy efficiency of the UHT plant. Over 90 per cent of the heat can be regenerated, although this figure varies with the type of plant. In some plants, the milk is held for some time in a holding tube after preheating, e.g., for 60 seconds at ~95°C as in figure. The major reason for this step is to reduce the amount of fouling, or deposit formation, in subsequent heat exchangers although, as noted below, it can also have a major effect on the quality of the final product by inactivating a natural milk enzyme.

The final heating step to the required sterilisation temperature is achieved by one of two major types of heating, the so-called direct and indirect systems. Direct systems heat milk by direct contact with culinary superheated steam while indirect systems employ heat exchangers in which superheated steam heats the milk indirectly through a stainless steel barrier in the form of either a tube or a plate. Direct systems can be either an injection type in which steam is injected into the milk, or an infusion type in which milk is infused into a chamber of superheated steam. The major difference between direct and indirect systems is the rate at which the milk is heated. Direct systems heat milk from preheat temperature to sterilisation temperature in less than one second whereas indirect systems can take several seconds to minutes. The major consequence of this difference is that, for the same bactericidal effect, the direct systems produce much less chemical change in the milk constituents than the indirect systems.

When the sterilisation temperature is reached, the milk enters a holding tube. The temperature of the milk and the time it takes to pass through this holding tube are the nominal conditions which are usually cited for a UHT process, e.g., 140°C for five seconds. While this is a convenient convention, it does not give a true picture of the heat process to which the product is subjected. Many chemical and microbiological changes occur in the heating step immediately before and in the cooling step immediately after the sterilisation step and hence these sections of the plant should be taken into account in addition to the sterilisation holding tube when considering the extent of these changes.

The initial cooling of the product in direct systems is achieved very rapidly as it is passed through a vacuum chamber which removes the water condensed into the product during the steam heating and in so doing returns the temperature of the product to close to the temperature from which it was heated, usually around 75°C. In the final cooling step in direct systems, and in both cooling steps in indirect systems, the heat from the hot milk is transferred to the cold milk in the preheating/heat regeneration steps.

When fat is present in the product, such as in whole milk, a homogenisation step is included. This is carried out at 60-70°C, either before or after the sterilisation step. If the homogeniser is downstream of the sterilisation step, it must be aseptic as no bacteria can be introduced after sterilisation. This clearly puts a high demand on the plant operators to ensure the homogeniser is aseptic and for this reason, where possible, homogenisation is carried out before sterilisation. However, it has been found that milk processed by a direct heating process has to be homogenised downstream to break up aggregates of protein which form during heating and cause an astringent taste in the milk.

The aseptic packaging step is a crucial one. The product must be transferred after cooling to the final package and the package sealed without introducing even one bacterial cell. In most commercial plants, the product is held in an aseptic tank before it is sent to the aseptic packer. Various packaging types are available but the most common are paperboard and multilayered plastic. The packages are sterilised before being filled, usually with hot hydrogen peroxide followed by hot air to remove residual peroxide.

Changes in Milk during UHT Processing

It is inevitable that heating a product such as milk at temperatures up to ~140°C will have some effect on its constituents, in addition to the intended bactericidal effects. Furthermore, storage at room temperature for long periods of time (up to 12 months) causes additional effects.

To consumers used to drinking pasteurised milk which differs little in flavour from raw milk, UHT milk often appears to have a cooked or heated flavour. Modern UHT technology minimises the production of this flavour but most consumers can still detect it and it is one reason why many consumers prefer pasteurised milk. The typical flavour of UHT milk is due a combination of flavours, the chief of which are sulphurous flavours caused by volatile sulphur compounds released from the whey protein, and the proteins in membrane surrounding the milk fat globule. Other contributors are the aliphatic carbonyl compounds formed during heating and compounds formed in the Maillard reaction. Immediately after manufacture, UHT milk has a strong sulphurous smell and taste due to hydrogen sulphide and other volatile sulphur compounds such as methane thiol. These compounds are markedly reduced in the first week, presumably through oxidation.

The initial step in the Maillard reaction is the reaction between lactose and lysine in milk proteins, chiefly whey proteins. In fact, the extent of this reaction is an indication of the intensity of heat given to the milk. In practice, it is measured as furosine, a product formed when the lactose-containing protein is subjected to acid hydrolysis. Another indicator of the heat treatment is lactulose, an isomer of lactose.

The whey proteins, particularly β-lactoglobulin which forms about 50 per cent of these soluble proteins in milk, are denatured by heating over about 70°C so that in UHT milk, a large percentage of the whey proteins are in the denatured state and exist largely as complexes with caseins.

The instability of the whey proteins to heat has another consequence during UHT processing. Some whey protein denatures and attaches to the surfaces of the heat exchangers in proteinaceous deposits which obstruct the flow of milk and can eventually cause the plant to be closed down for cleaning. However, this is not the only type of deposit formed during UHT processing. At high temperatures, above about 110°C, calcium phosphate also precipitates on the walls, adding to the 'fouling' caused by the whey proteins.

Surprisingly, the UHT process has only a minimal effect on the nutrient value of milk. There is a small decrease in the water-soluble vitamins but virtually no change in the fatsoluble vitamins. The proteins, in fact, have been shown to be more digestible in UHT milk as a result of the heat treatment. UHT treatment may also reduce the allergenicity of the milk proteins.

The chemical changes caused by a particular plant can be summarised in a chemical index, C^*. A C^* of 1 is equivalent to three per cent destruction of the B vitamin, thiamine. UHT plants should be run under conditions which give a C^* of less than 1 to avoid excessive chemical damage. Direct heating UHT systems have lower C^* values than indirect systems. Therefore a better description of a UHT plant is provided by its B^* and C^* rather than by the conventionally used temperature-time combination of the sterilisation holding tube.

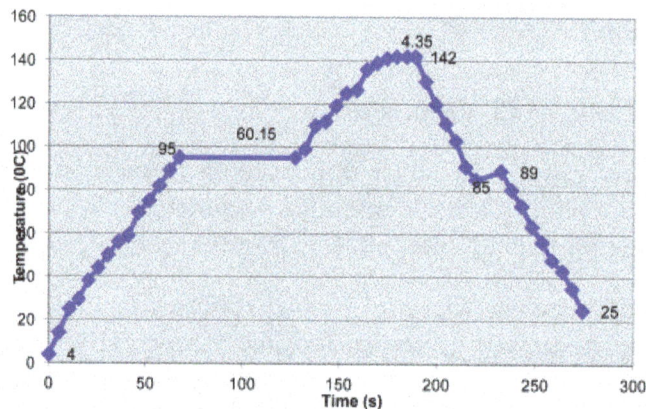

Temperature-time profile of a commercial indirect-heating UHT plant. The nominal holding tube conditions for this plant are 142°C for 4.35 seconds and it's B^* and C^* values are 5.77 and 1.52 respectively

Most of the above changes that occur during high-temperature processing of milk have been studied in depth and their reaction kinetics worked out. These enable the changes occurring in a particular UHT plant to be mathematically estimated. The basic information on a plant which allows this to be done is the temperature-time profile. This profile can vary considerably as illustrated in Figures, which show profiles of two commercial UHT installations, an indirect and a direct system, respectively. Therefore when the kinetics for a particular change, e.g., denaturation of β-lactoglobulin, and the temperature-time profile of a UHT plant are known, the effect of that plant on a range of milk components can be predicted. Fortunately, the calculations for doing this can be performed by a computer and software for this purpose is available commercially, e.g. NIZO Premia. The B^* and C^* values mentioned above can also be computed as they are defined by mathematical formulae. Unfortunately, it is not easy to obtain the temperature-time profile for most plants as a full description of the plant in terms of temperatures and times in all sections is not readily available. Where they are available however, a wealth of information about the overall plant and the product produced it can be obtained. One application of this computer simulation is comparing the effects of small-scale pilot plants with full-sized commercial plants to enable the pilot plants to be configured to match the full-sized commercial UHT equipment.

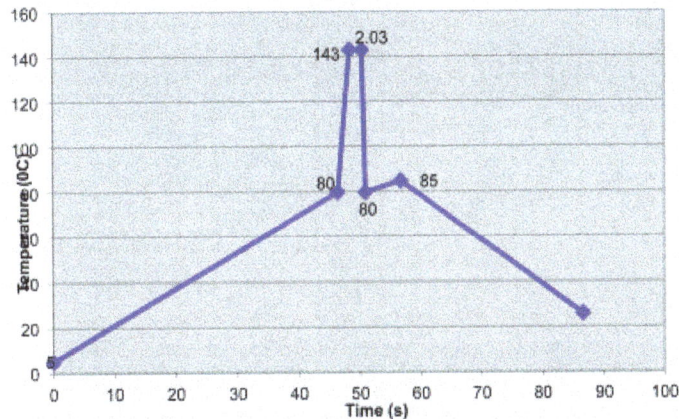

Temperature-time profile of a commercial direct-heating UHT plant. The nominal holding tube conditions for this plant are 143°C for 2.03 seconds, and it's B* and C* values are 1.26 and 0.17 respectively

Changes in UHT Milk during Storage

Keeping milk in good condition at room temperature for up to 12 months is a major challenge because of the myriad of changes which can take place. The flavour changes through progress of the Maillard reaction and through oxidation by dissolved oxygen in the milk. The major flavour compounds produced are methyl ketones and aliphatic aldehydes but a large number of flavour compounds are generated. Other flavours which may develop during storage are due to the action of heat-resistant bacterial enzymes which may be present in the raw milk and survive the UHT heat treatment. These include lipases, which break down the fat and form free fatty acids, some of which have strong flavours, and proteases which break down proteins to produce peptides, some of which are bitter.

Another change which can be brought about by proteases is what is known as 'age gelation' where the milk thickens over storage and eventually turns into a gel akin to a yogurt. This undesirable defect can be caused by the heat-resistant bacterial enzymes but it can be also caused by plasmin, a naturally occurring protease in milk, which is quite heat stable and can remain active in UHT milk. Recently, it has been found to be inactivated by some UHT preheating conditions which is an excellent reason for including a holding time in the preheat section of UHT plants.

UHT Processing of Products Other than White Milk

UHT processing is now widely used for producing 'long life' products such as cream, custard and flavoured milks. However, it is not suitable for making cheese as the curd from UHT milk takes a long time to set and retains a high amount of moisture, giving a very soft and unacceptable cheese. UHT milk is also not very suitable for yogurt manufacture as it forms a very soft gel; however, it may be more suitable for producing a (long life) drinking yogurt where a firm gel is not required.

Dry Milling and Fractionation of Grain

Milling transforms cereals into more-palatable, more-desirable food ingredients. Dry milling is the separation of the anatomical parts of the grain as cleanly as possible. Subsequently, some of the parts are reduced in particle size. Milling generally involves recovery of the main tissue (i.e., the starchy endosperm) and the concomitant removal of the material the miller calls "bran" (i.e., the pericarp, the seed coat, the nucellar epidermis, and the aleurone layer). In addition, the germ is usually removed from the endosperm. Because of the relatively high oil content of the germ, its presence increases the risk of the product becoming rancid and thereby less palatable. The bran and germ are relatively rich in protein, dietary fiber, B vitamins, minerals, and fat, and the separated endosperm is therefore lower in these components than the original grain. Thus, while milling increases the palatability of cereal products, it decreases the nutritional value of the main product obtained.

In dry milling, the particle size distribution of the endosperm products obtained is dictated by the end use of the product. In general, it is desirable for rice or barley endosperm to remain in one large piece, while a large grit is desirable from maize and coarse semolina from durum wheat. Wheat and rye, at opposite ends of the size spectrum, are generally milled into fine flour products. It follows that different types of equipment are used for dry milling. However, they consistently aim to produce palatable products with a good shelf life.

Features

The dry milling process includes a number of unique features:

- Physical separation (size/density) based on mass

- No use of chemicals

- Maximizing surface area of solids for processing

- Resource pulping

- Minimal water use, if any (short tempering)

- Note: Water is not used as a separation agent

- Low capital cost

- Lower separation compared to wet milling

- Lower concentration of starch, protein, fiber, and oil relative to wet milling

The most utilized grinding mills include pin, hammer, and disk mills, but many machines are utilized for more specific processes. To maintain a high starch extraction, the grains will go through a degermination process. This process removes the germ and fiber (pericarp) first, and the endosperm is recovered in several sizes: grits, cones, meal, and flow. It is important to note that the gluten protein matrix is not separated from the starch.

Yields

The table below is a compilation of particle size and yield of milled maize products.

Product	Particle size range (mm)	Yield (% by weight)
Flaking grits	5.8-3.4	12
Coarse grits	2.0-1.4	15
Medium grits	1.4-1.0	23
Fine grits	1.0-0.65	23
Coarse meal	0.65-0.3	10
Fine meal	0.3-0.17	10
Flour	<0.17	5

Uses

Currently, products of dry milled corn products are used mostly in animal food, brewing and breakfast cereals industries.

Grits/Cones

- Breakfast cereals

- Snack foods
- Pet foods
- Corn bread
- Breads

Flour

- Baby foods
- Baking mixes
- Batters
- Desserts
- Frozen foods
- Meat extenders
- Thickening agents

Germ

- Grain based oil
- Vitamin carriers
- Mayonnaise
- Potato chips
- Soups
- Sauces
- Livestock feed

During alcohol production, the main advantage of dry milling is the flexibility in type and quality of grain which can be utilized as substrates for the fermentation process. Dry milling can be utilized for a number of different grains with little to no alteration to machine operation characteristics.

Laws of Grinding

Currently, there are three main empirical models which are used to calculate the grinding work required relative to grain size and quantity. The Kick model may be utilized for grains with diameters greater than 50 mm; the Bond model for grain diameter between 0.05 mm – 50 mm; the Von Rittinger model for grain less than 0.05 mm. The calculations are shown here:

Kick Model

$$W_k = c_k (\ln d_A - \ln d_E)$$

Bond Model

$$W_B = c_b \left(\frac{1}{\sqrt{d_E}} - \frac{1}{\sqrt{d_A}} \right)$$

Von Rittinger Model

$$W_R = c_R \left(\frac{1}{d_E} - \frac{1}{d_A} \right)$$

In all three models:

W is the work in kj/kg;

c is the grinding coefficient;

d_A is the grain size of the source grain;

d_E is the size of the ground material.

While Bond's coefficient may be viewed in various literature, the calculation for Kick's and Von Rittinger's coefficients may be viewed below:

$$c_k = 1.151 \frac{c_B}{\sqrt{d_{BU}}}$$

$$c_R = 0.5 \frac{c_B}{\sqrt{d_{BL}}}$$

where, d_{BU} = 50mm and d_{BL} = 0.05mm.

To analyze the grinding results the dispositions of the source and ground material must be computed. To quantify this characteristic the grinding degree is calculated, which is a ratio of sizes relative to the grain disposition. The grinding degree relative to grain size d_{80} is shown:

$$Z_d = \frac{d_{80,1}}{d_{80,2}}$$

where, the d_{80} value signifies 80% mass is of size smaller than the grain.

Types of Dry Grinders

There are three methods used for corn dry milling:

- Alkaline-cooked process
- Stone-ground (non-degerming process)
- Tempering degerming process

Tempering degerming is the most common process used for grain dry milling in the industry.

Process Overview

Corn dry milling consists of several steps. The following paragraphs describe all the steps of dry milling as well as the equipment used during these steps in detail.

Tempering

A chamber is used at this section in order to mix the corn and water and let them temper for 10 to 30 minutes. For more efficient separation, differential moisture content between germ and endosperm is desired. Tempering of kernel leads to moisture uptake. Because of the differential swelling of germ and endosperm, the germ becomes more flexible and resilient during tempering while there is no movement of material out of kernel.

Degermination

The objective of degermination in corn dry milling is to break down kernel to pericarp, endosperm and germ. Beall operation is used for fulfilling this goal which separates the kernels received form tempering section into tails and throughs. Beall degerminator is known for its high yield of flaking grits; however, other manufactures have lower power requirement. Pilot plant Beall has an inner cone rotating at 800 rpm. One of the advantages of Beall degermination is weight adjustment at tailgate for increment the residence time by holding back the material.

Aspiration

Aspiration is a unit operation used for separating the pericarp from the mixture of endosperm and germ, using terminal velocity which is affected by particle size, shape and density.

Gravity Separation

Gravity separation is a method used for separating the components of a mixture with different specific weight. Gravity separation in dry milling is utilized in order to separate endosperm from germ.

Roller Milling and Sifting

Roller mills use cylindrical rollers for grinding different materials, especially grains, which can even be an appropriate substitution for hammer mill and ball mills. Sifting is used in order to adjust the distribution of endosperm particle

Oil Recovery

There are two methods for oil recovery used in industry:

1) Corn expelling which is not commercially used in the US for corn germ oil recovery due to its low oil yield and presence of residuals oil in solid products; however, it is simple and cheap.

2) Extraction which is mostly used because of its high oil yield and lower residual oil although it is expensive and has explosion and safety risks.

Ohmic Heating

Ohmic heating is also known as joule heating, electric resistance heating, direct electric heating, electro heating and electro conductive heating. It is a process in which alternating electric current is passed through food material to heat them. Heat is internally generated within the material owing to the applied electrical current. In conventional heating, heat transfer occurs from a heated surface to the product interior by the means of convection and conduction and is time consuming especially with longer conduction or convection paths that may exist in the heating process. Elecroresistive or ohmic heating is volumetric in nature and thus has the potential to reduce over processing by virtue of its inside-outside heat transfer pattern. Ohmic heating is distinguished from other electrical heating method by the presence of electrodes contacting the food by frequency or by waveform.

Ohmic heating is not a new technology; it was used as a commercial process in the early twentieth century for the pasteurization of milk. However, the electro pure process was discontinued between the late 1930s and 1960s ostensibly because of the prohibitive cost of the electricity and a lack of suitable electrode material.

Interest in ohmic heating was rekindled in the 1980s, when investigators were searching for viable methods to effectively sterilize liquid- large particle mixtures, a scenario for which aseptic processing alone was unsatisfactory.

Ohmic heating is one of the newest methods of heating foods. It is often desirable to heat foods in a continuous system such as heat exchanger rather than in batches as in a kettle or after sealing in a can. Continuous systems have the advantage that they produce less heat damage in the product, are more efficient, and they can be coupled to aseptic packaging systems. Continuous heating systems for fluid foods that contain small particles have been available for many years. However, it is much more difficult to safely heat liquids containing larger particles of food. This is because it is very difficult to determine if a given particle of food has received sufficient heat to be commercially sterile. This is especially critical for low acid foods such as Beef stew which might cause fatal food poisoning if under heated. Products tend to become over processed if conventional heat exchangers are used to add sufficient heat to particulate foods. This concern has hindered the development of aseptic packaging for foods containing particulates. Ohmic heating may over come some of these difficulties and limitations.

Considerable heat is generated when an alternating electric current is passed through a conducting solution such as a salt brine. In ohmic heating a low-frequency alternating current of 50 or 60 Hz is combined with special electrodes. Products in a conducting solution (nearly all polar food liquids are conductors) are continuously passed between these electrodes. In most cases the product is passed between several sets of electrodes, each of which raise the temperature. After heating, products can be cooled in a continuous heat exchanger and then aseptically filled into presterlized containers in a manner similar to conventional aseptic packaging. Both high and low- acid products can be processed by this method.

An advancement in the thermal processing is ohmic heating. In principle, electricenegy is transformed into thermal energy uniformly throughout the product. Rapid heating results, and better nutritional and organoleptic qualities are possible when compared with conventional in -can sterilization. "Ohmic heating employs electrodes immersed on pipe," Quass says. " Product is pumped through the pipe as current flows between the electrodes." Depth of penetration is not limited. The extent of heating is determined by the electrical conductivity through the product, plus residence time in the electric field. "ohmic heating is useful for foods thus burn-on or have particulates that plug up heat exchangers," continues Quass. "Instead of using a scraped surface heat exchanger for stew, for example, ohmic heating can reduce the come-up time, and improve product quality.

Ohmic heating is defined as a process wherein (primarily alternating) electric currents are passed through foods or other materials with the primary purpose of heating them. The heating occurs in the form of internal energy generation within the material. Ohmic heating is distinguished from other electrical heating methods either by the presence of electrodes contacting the food (as opposed to microwave and inductive heating, where electrodes are absent), frequency (un-restricted, except for the specially assigned radio or microwave frequency range), and waveform (also unrestricted, although typically sinusoidal).In inductive heating, electric coils placed near the food product generate oscillating electromagnetic fields that send electric currents through the food, again primarily to heat it. Such fields may be generated in various ways, including the use of the flowing food material as the secondary coil of a transformer. Inductive heating may be distinguished from microwave heating by the frequency (specifically assigned in the case of micro-waves), and the nature of the source (the need for coils and magnets for generation of the field, in the case of inductive heating, and a magnetron for microwave heating).Information on inductive heating is extremely limited.

A project was conducted in the mid-1990s at the Technical University of Munich (Rosenbauer 1997), under sponsorship from the Electric Power Research Institute. No data about microbial death kinetics under inductive heating were published. Thus, the succeeding discussion focuses on ohmic heating. A large number of potential future applications exist for ohmic heating, including its use in blanching, evaporation, dehydration, fermentation, and extraction. The present discus-sion, however, concerns primarily its application as a heat treatment for microbial control. In this sense, the main advantages claimed for ohmic heating are rapid and relatively uniform heating. Ohmic heating is currently being used for processing of whole fruits in Japan and the United King-dom. One commercial facility in the United States uses ohmic heating for the processing of liquid egg. The principal advantage claimed for ohmic heating is its ability to heat materials rapidly and uniformly, including products containing particulates. This is expected to reduce the total thermal abuse to the product in comparison to conventional heating, where time must be allowed for heat penetration to occur to the center of a material and particulates heat slower than the fluid phase of a food. In ohmic heating, particles can be made to heat faster than fluids by appropriately for-mulating the ionic contents of the fluid and particulate phase to ensure the appropriate levels of electrical conductivity.

Principle of Ohmic Heating

Ohm's law states that, at constant temperature in an electrical circuit, the current passing through a conductor between two points is directly proportional to the potential difference (i.e. voltage drop or voltage) across the two points, and inversely proportional to the resistance between them.

The mathematical equation that describes this relationship is:

$$I = V/R$$

Where,

I is the current in amperes, V is the potential difference between two points of interest in volts, and R is a circuit parameter, measured in ohms (which is equivalent to volts per ampere), and is called the resistance. The potential difference is also known as the voltage drop, and is sometimes denoted by U, E or emf (electromotive force) instead of V.

The law was named after the physicist Georg Ohm, who, in a treatise published in 1827, described measurements of applied voltage and current passing through simple electrical circuits containing various lengths of wire. He presented a slightly more complex equation than the one above to explain his experimental results (the above equation is the modern form of Ohm's law; it could not exist until the ohm itself was defined (1861, 1864)). Well before Georg Ohm's work, Henry Cavendish found experimentally (January 1781) that current varies in direct proportion to applied voltage, but he did not communicate his results to other scientists at the time.

The resistance of most resistive devices (resistors) is constant over a large range of values of current and voltage. When a resistor is used under these conditions, the resistor is referred to as an ohmic device because a single value for the resistance suffices to describe the resistive behavior of the device over the range. When sufficiently high voltages are applied to a resistor, forcing a high current to flow through it, the device is no longer ohmic because its resistance, when measured under such electrically stressed conditions, is different (typically greater) from the value measured under standard conditions.

Ohm's law, in the form above, is an extremely useful equation in the field of electrical/electronic engineering because it describes how voltage, current and resistance are interrelated on a macroscopic level, that is, commonly, as circuit elements in an electrical circuit.

Parameters of Importance in Ohmic Heating

Product Properties

The most important parameter of interest in ohmic heating is the electrical conductivity of the food and food mixture. Substantial research was conducted on this property in the early 1990s because of the importance of electrical conductivity with regard to heat transfer rate and temperature distribution. The electrical conductivity is determined using the following equation:

$$Đ' = L / AR$$

Where Đ' is the specific electrical conductivity (S/m), A the area of cross section of the sample (m2), L the length of the sample (m), and R the resistance of the sample (ohm). General findings of numerous electrical conductivity studies are as follows.

The electrical conductivity is a function of food components; ionic components (salt), acid, and moisture mobility increase electrical conductivity, while fats, lipids, alcohol decrease it. Electrical conductivity is linearly correlated with temperature when the electrical field is sufficiently high (at least 60 V/cm).

Nonlinearities (sigmoid curves) are observed with lower electrical field strength.

Electrical conductivity increases as the temperature and applied voltage increases and decreases as solids content increases.

Lowering the frequency of AC during ohmic heating increases the electrical conductivity.

The waveform can influence the electrical conductivity; through AC is usually delivered in sine waves, sawtooth waves increased the electrical conductivity in the some cases, while square waves decreased it.

Electrical conductivity as opposed to raw sample showed increased electrical conductivity as opposed to raw samples when both were subsequently subjected to ohmic heating.

The electrical conductivity of solids and liquids during ohmic heating of multiphase mixtures is also critically important. In an ideal situation, liquid and solid phases posses essentially equal electrical conductivities and would thus (generally) heat at the same rate. When there are differences in the electrical conductivity between a fluid and solid particles, the particles heat more slowly then a fluid when the electrical conductivity of the solid is higher than that of the fluid. Fluid motion (convective heat transfer) is also an important consideration when there are electrical conductivity differences between fluids and particles.

Other product properties that may affect temperature distribution include the density and specific heat of the food product. When solid particles and a fluid medium have similar electrical conductivities, the component with the lower heat capacity will tend to heat faster. Heat densities and specific heats are conductive to slower heating. Fluid viscosity also influences ohmic heating; higher viscosity fluids tend to result in faster ohmic heating than lower viscosity fluids.

Texture Analysis

Sensory evaluation is critically important to any viable food processes. Numerous publications have cited the superior product quality that can be obtained through decreased process time, though few published studies specifically quantify sensory and texture issues. Six stew formulations sterilized using ohmic heating before and after 3 years of storage were analyzed; the color, appearance, flavor, texture, and overall food quality ratings were excellent. 'Indicating that ohmic heating technology has the potential to provide shelf-stable foods mechanical properties of hamburgers cooked with a combination of conventional and ohmic heating were not different from hamburgers cooked with conventional heating.

Microbial Death Kinetics

In terms of microbial death kinetics, considerable attention has been paid to the following question: does electricity result in microbial death, or is microbial death caused solely by heat treatment? The challenge in modeling microbial death kinetics is precise matching of time-temperature histories between ohmic heating and conventional process. The FDA has published a comprehensive review of microbial death kinetics data regarding ohmic heating.

Initial studies in this area showed mixed results, though the experimental details were judged insufficient to draw meaningful conclusions. Researches compared death kinetics of yeast cells under ohmic heating. More recent work in this area has indicated those decimal reduction times of Bacillus Subtiles spores were significantly reduced when using ohmic heating at identical temperatures. These investigators also used a two-step treatment process involving ohmic heating, followed by holding and heat treatment, which accelerated microbial death kinetics. The inactivation of yeast cells in phosphate buffer by low-amperage direct current (DC) electrical treatment and conventional heating at isothermal temperature was examined. These researchers concluded that a synergistic effect of temperature and electrolysis was observed when the temperature became lethal for yeast.

Future research regarding microbial death kinetics, survivor counts subsequent to treatment, and the influence of electricity on cell death kinetics are necessary to address regulatory issues. At the present time, assuming that microbial death is only a function of temperature (heat) results in an appropriately conservative design assumption.

Vitamin Degradation Kinetics

Limited information exists regarding product degradation kinetics during ohmic heating. Researchers measured vitamin C degradation in orange juice during ohmic and conventional heating under nearly identical time-temperature histories and concluded that electricity did not influence vitamin C degradation kinetics. This study was conducted at one electrical field strength (E=23.9 V/cm). Others found that the ascorbic acid degradation rate in buffer solution during ohmic heating was a function of power, temperature, NaCl concentration, and products of electrolysis. Further research in this area could include the influence of electrical field strength, end point temperature and frequency of AC on the degradation of food components during ohmic heating. The characterization of electrolysis is also critical need in this area.

Mechanisms of Microbial Inactivation

The principal mechanisms of microbial inactivation in ohmic heating are thermal in nature. Occasionally, one may wish to reduce the process requirement or to use ohmic heating for a mild process, such as pasteurization. It may then be advantageous to identify additional non-thermal mechanisms. Early literature is inconclusive, since temperature had not been completely eliminated as a variable. Recent literature that has eliminated thermal differences, however, indicates that a mild electroporation mechanism may occur during ohmic heating. The principal reason for the additional effect of ohmic treatment may be its low frequency (50 - 60 Hz), which allows cell walls to build up charges and form pores. This is in contrast to high-frequency methods such as radio or microwave frequency heating, where the electric field is essentially reversed before sufficient charge buildup occurs at the cell walls.

Applications of Ohmic Heating in Food Industries

Ohmic heating can be applied to wide variety of foods, including liquids, solids and fluid-solid mixture.

Ohmic heating is being used commercially to produce liquid egg products in United States.

It is being used in the United Kingdom and Japan for the processing of whole fruits such as Straw-berries.

Additionally, ohmic heating has been successfully applied to wide variety of foods in lab including Fruits and Vegetables, juices, sauces, stew, meats, seafood, pasta and soups.

Widespread commercial adoption of ohmic heating in the United states is dependent on regulatory approval by the FDA, a scenario that requires full understanding of the ohmic heating process with regard to heat transfer (temperature distribution), mass transfer (concentration distribution, which are influenced by electricity), momentum transfer (fluid flow) and kinetic phenomena (thermal and possibly electro thermal death kinetics and nutrient degradation).

Effect of Ohmic Heating on Food Products

Products

1. Ohmic heating could up juice quality:

Israeli scientists say that ohmic heating of orange juice has proved to be good way of improving the flavor quality of orange juice while extending sensory shelf life.

The scientists were observed that sensory shelf life of orange juice could be extended to more than 100 days, doubling expectancy compared to pasteurization methods. Ohmic heating uses electricity to rapidly and uniformly heat food and drink, resulting in less thermal damage to the product. The technology has been around since the early 1900s, but it was not until the 1980s that food processing researchers began investigating the possible benefits to the industry.

The scientists compared pasteurized orange juice, which had been heated at 90Ëšc for 50 sec, with orange that was treated at 90,120 and 150Ëšc for 1.13, 0.85 and 0.68 sec in an ohmic heating system. The experiment found that for all examples retention of both pectin and vit. C was reported similar. Likewise both treatments prevented the growth of micro-organisms for 105 days, compared to fresh orange juice. However, where the ohmic heated samples proved much stronger was in the preservation of flavors and the general taste quality over a period of time. The scientists tested five representative flavor compounds- decanal, octanol, limonene, pinene and mycrene. Testing showed that levels of these compounds were significantly higher in the ohmic treated samples after storage than in the pasteurized examples.

The scientists' results found that only adverse reaction that the ohmic treated orange juice had that it increased browning in the juice, although this was not reported to be visible, until after 100 days. Conversely the appearance of the ohmic heated samples was said to be visibly less cloudy. The implications of the findings to the juice industry could be wide reaching as quality is a major driving force for a product that is often marketed in the premium category. If the cost of implementation proves competitive then this could become a serious contender to pasteurized methods.

2. Ohmic heating behavior of hydrocolloid solutions:

Aqueous solutions of five hydrocolloids (Carrageenan, 1-3%; xanthan, 1-3%; pectin, 1-5%; gelatin, 2-4% and starch, 4-6%) were heated in a static ohmic heating call at a voltage gradient of 7.24V cm-1. Time and temperature data, recorded at selected time intervals, were used to study the effect of concentration and temperature on the ohmic heating behavior of hydrocolloid solutions. Of the test samples examined, carrageenan gave the shortest time to raise the temperature from 20 to 100Ëšc: 4200,1600 and 1100s at 1, 2 and 3% concentration respectively. For the same temperature raise, xanthan samples required 5500, 2300 and 1400s at 1, 2 and 3% concentration levels. Pectin and gelatin samples were found to exhibit even lower, but similar heating profiles. At highest concentration (5%), pectin took 7300s to reach 100 from 20Ëšc, and at all other concentrations, the time limit of 10,000s was exceeded before it reached 100Ëšc. The temperature of starch solutions never exceeded 62Ëšc within the specified time limit. Heating was found to be uniform throughout samples for carrageenan, pectin (1-3%) and gelatin samples. For xanthan and starch solutions, some non-uniformity in temperature profiles was observed. The observed ohmic heating behavior of hydrocolloid solutions corresponded well with their electrical conductivity values. The homogenesity of heating was related to rheological properties of hydrocolloid solutions and values. The homogenesity of heating was related to rheological properties of hydrocolloid solutions and their behavior at high temperature.

3. Design and performance evaluation of an ohmic heating unit for liquid foods:

An experimental ohmic heating unit was designed and fabricated for continuous thermal processing of liquid foods. The unit was supported by a data acquisition system for sensing the liquid temperature distribution, line voltage and current with time. A separate small ohmic heating unit was also used for batch heating tests. The data acquisition system performed well and could record temperatures, voltage and current at intervals of two seconds. The performance of the ohmic heating unit was evaluated based on batch and steady state continuous flow experiments. Tests with 0.1 M aqueous sodium chloride solution showed the ohmic heating to be fast and uniform. In batch heating tests, the electrical conductivity of the liquid could be determined easily as a function of temperature using instantaneous values of the voltage gradient and current density. In continuous flow heating experiments, other physical properties, applied voltage gradient and dimensions of unit the heating.

4. Determination of starch gelatinization temperature by ohmic heating:

A method for measuring starch gelatinization temperature (T), determined from a change in electrical conductivity (Ð±), was developed. Suspension of native starches with different starch/ water mass ratios and pre-gelatinized starches were prepared, and ohmicallly heated with agitation to 90Ëšc using 100V by AC power at 50 Hz, and a voltage gradient of 10 V/cm. the results showed that Ð± of native starch suspensions was linear with temperature expect for the gelatinization range, but the linear relationship was always present for the pre-gelatinized starch-water system. It was seen that the shape of dÐ±/dT versus T curve was essentially similar to the endothermic peak on a DSC thermo gram, and the gelatinization temperature could be conveniently determined from this curve. Thus, the segment profile on this curve was called the "block peak". The reason for the decrease in Ð± of native starch suspension in the gelatinization range was probably that the area foe motion of the charged particles was reduced by the swelling of stearch granules during gelatinization.

5. Ohmic heating of strawberry products: electrical conductivity measurements and ascorbic acid degradation kinetics:

The effect of field strength and multiple thermal treatments on electrical conductivity of strawberry products were investigated. Electrical conductivity increase with temperature for all the products and conditions tested following linear relations. Electrical conductivity was found to depend on the strawberry- based product., an increase of electrical conductivity with field strength was obvious for two strawberry pulps and strawberry filling but not for strawberry-apple sauce. Thermal treatments caused visible changes (a decrease) in electrical conductivity values of both strawberry pulps tested, but the use of a conventional or ohmic pre-treatment induces a different behavior of the pulps' conductivity values. Ascorbic acid degradation followed first order kinetics for both conventional and ohmic heating treatments and the kinetic constants obtained were in the range of the values reported in the literature for other food systems. The presence of an electric field does not affect ascorbic acid degradation.

6. Polyphenoloxidase deactivation kinetics during ohmic heating of grape juice:

The heating method affects the temperature distribution inside a food and directly modifies the time-temperature relationship for enzyme deactivation. Fresh grape juice was ohmically heated at different voltage gradient (20, 30 and 40 V/cm) from 20EsC to temperatures of 60, 70, 80 or 90Ëšc and the change in the activity of polyphenoloxidase enzyme (PPO) was measured. The critical deactivation temperatures were found to be 60Ëšc or lower for 40V/cm were fitted to the experimental data. The simplest kinetic model involving one step first-order deactivation was better than more complex models. The activation energy of the PPO deactivation for the temperature range of 70-90Esc was found to be 83.5 kJ/mol.

7. Processing and stabilization of cauliflower by ohmic heating technology:

Cauliflower is a brittle product which does not resist conventional thermal treatments by heat. The feasibility of processing cauliflower by ohmic heating was investigated. Cauliflower florates were sterilized in 10 kW APV continuous ohmic heating pilot plant with various configurations of pre-treatments and processing conditions. The stability of final products was examined and textural qualities were evaluated by mechanical measurements. Ohmic heating treatments gave a product of attractive appearance, with interesting firmness properties and proportion of particles >1cm. stabilities at 25Ëšc and 37Ëšc were verified and in one case, the product was even stable at 55Ëšc. Low temperature precooking of cauliflower, high rate and sufficient electrical conductivity of florates seem to be optimal conditions. The interest of using this electrical technology to process brittle products such as ready meals containing cauliflower was high lightened.

8. Electrical conductivity of apple and sour cherry juice concentrates during ohmic heating:

Ohmic heating is based on the passage of electrical current through a food product that serves as an electrical resistance. In this study, apple and sourcherry concentrates having 20-60% soluble solids were ohmically heated by applying five different voltage gradients (20-60 V/cm). The electrical conductivity relations depending on temperature, voltage gradient and concentration were obtained. It was observed that the electrical conductivities of apple and sourcherry juices were significantly affected by temperature and concentration ($P < 0.05$). The ohmic heating system performance coefficients (SPCs) were defined by using the energies given to the system and taken up by the juice samples. The SPCs were in the range of 0.47-0.92. The unsteady-state heat conduction equation for negligible internal resistance was solved with an ohmic heating generation term by the finite difference technique. The mathematical model results considering system performance coefficients were compared with experimental ones. The predictions of the mathematical model using obtained electrical conductivity equations were found to be very accurate.

The Commercial Development of Ohmic Heating Processes

The problems of heat transfer techniques in cook-chill food processing. This include destruction of flavours and nutrients, and particle damage arising from high shear often employed to improve heat transfer rates. These heat transfer problems have now been overcome with the development of ohmic heating technology. The ohmic heating effect occurs when an electric current is passed through an electrically conducting product. In practice, low frequency alternating current (50 or 60 Hz) from the public mains supply is used to eliminate the possibility of adverse electro-chemical reactions and minimise power supply complexity and cost. Electrical energy is transformed into thermal energy. The depth of penetration is virtually unlimited and the extent of heating is governed only by the spacial uniformity of electrical conductivity throughout the product and its residence time in the heater. The authors briefly discuss the design features, temperature control and market acceptance of ohmic heating.

Advantages of Ohmic Heating

Ohmic heating exhibits several advantages with respect to conventional food processing technologies as follows:

- Particulate foods upto 1 in are suitable for ohmic heating; the flow of a liquid particle mixture approaches plug flow when the solids content is considerable (20-70%).

- Liquid particle mixtures can heat uniformly under some circumstances (for example, if liquids and particles posses similar electrical conductivities or if properties such as solids concentration, viscosity, conductivity, specific heat and flow rate are manipulated appropriately).

- Temperatures sufficient for ultra high temperature (UHT) processing can be rapidly achieved.

- There are no heat surfaces for heat transfer, resulting in a low risk of product damage from burning or over processing.

- Energy conversion efficiencies.

- Relatively low capital cost.

References

- Jane E Henney; Christine L Taylor; Caitlin S Boon, eds. (2010). Strategies to Reduce Sodium Intake in the United States. Washington, D.C.: National Academies Press, National Academy of Sciences. ISBN 978-0-309-14805-4. PMID 21210559

- All-about-brining-331490: thespruceeats.com, Retrieved 21 March 2018

- Harper, J.M. (1978). "Food extrusion". Critical Reviews in Food Science and Nutrition. 11 (2): 155–215. doi:10.1080/10408397909527262. PMID 378548

- 2247-an-introduction-to-the-drying-food-preservation-method: homepreservingbible.com, Retrieved 29 May 2018

- Akdogan, Hülya (June 1999). "High moisture food extrusion". International Journal of food Science & Technology. 34 (3): 195–207. doi:10.1046/j.1365-2621.1999.00256.x

- Everything-about-food-extrusion: discoverfoodtech.com, Retrieved 19 July 2018

- Karwe, Mukund V. (2008). "Food extrusion". Food Engineering. 3. Oxford Eolss Publishers Co Ltd. ISBN 978-1-84826-946-0

- Mass-production-definition-techniques-examples: study.com, Retrieved 09 July 2018

- Camire, M.E. (1998). "Chemical changes during extrusion cooking. Recent advances". Advances in Experimental Medicine and Biology. 434: 109–121. doi:10.1007/978-1-4899-1925-0_11. PMID 9598195

- How-to-preserve-vegetables-by-snap-freezing: cookingforbusymums.com, Retrieved 15 June 2018

- Heldman, Dennis R.; Hartel, Richard W. (1997). Principles of Food Processing. Springer. ISBN 9780834212695

- Use-of-ohmic-heating-in-food-preservation-biology-essay: ukessays.com, Retrieved 25 June 2018

- Shivendra Singh, Shirani Gamlath, Lara Wakeling (10 May 2007). "Nutritional aspects of food extrusion: a review". International Journal of Food Science & Technology. 12: 916–929. doi:10.1111/j.1365-2621.2006.01309.x

Permissions

We would like to thank the editorial team for lending their expertise to make the book truly unique. They have played a crucial role in the development of this book. Without their invaluable contributions this book wouldn't have been possible. They have made vital efforts to compile up to date information on the varied aspects of this subject to make this book a valuable addition to the collection of many professionals and students.

This book was conceptualized with the vision of imparting up-to-date and integrated information in this field. To ensure the same, a matchless editorial board was set up. Every individual on the board went through rigorous rounds of assessment to prove their worth. After which they invested a large part of their time researching and compiling the most relevant data for our readers.

The editorial board has been involved in producing this book since its inception. They have spent rigorous hours researching and exploring the diverse topics which have resulted in the successful publishing of this book. They have passed on their knowledge of decades through this book. To expedite this challenging task, the publisher supported the team at every step. A small team of assistant editors was also appointed to further simplify the editing procedure and attain best results for the readers.

Apart from the editorial board, the designing team has also invested a significant amount of their time in understanding the subject and creating the most relevant covers. They scrutinized every image to scout for the most suitable representation of the subject and create an appropriate cover for the book.

The publishing team has been an ardent support to the editorial, designing and production team. Their endless efforts to recruit the best for this project, has resulted in the accomplishment of this book. They are a veteran in the field of academics and their pool of knowledge is as vast as their experience in printing. Their expertise and guidance has proved useful at every step. Their uncompromising quality standards have made this book an exceptional effort. Their encouragement from time to time has been an inspiration for everyone.

The publisher and the editorial board hope that this book will prove to be a valuable piece of knowledge for students, practitioners and scholars across the globe.

Index

www.ingramcontent.com/pod-product-compliance
Lightning Source LLC
Chambersburg PA
CBHW082035190326
41458CB00010B/3372